Makers:

All Kinds of People Making Amazing Things In Garages, Basements, and Backyards

By Bob Parks

From the Makers of MAKE Magazine

Beijing | Cambridge | Köln | London | Paris | Sebastopol | Taipei | Tokyo

Makers: All Kinds of People Making Amazing Things in Garages, Basements, and Backyards
by Bob Parks

ISBN: 0596101880

Printed in Canada.

Make: Books
O'Reilly Media, Inc.
1005 Gravenstein Highway North
Sebastopol, CA 95472

Executive Editor: Dale Dougherty
Editor: Mark Frauenfelder
Production Editors: Shawn Connally, Keith Hammond
Designed by: Albertson Design, San Francisco
Designers: David Albertson, Sara Huston

O'Reilly books may be purchased for educational, business, or sales promotional use. For more information, contact our corporate/institutional sales department: (800) 998-9938 or corporate@oreilly.com.

Printing History: January 2006: First Edition

ACKNOWLEDGEMENTS

>> I was hanging sheetrock — and taking a short break from reporting — when Dale Dougherty called with an amazing idea to profile 100 people who make things. I couldn't believe my luck; I dropped the taping knife and got started. I want to thank him for his vision and enthusiasm for this book. Turns out we tallied a mere 91 makers, not 100 — but the work was aided immensely by Dale's constant encouragement and that of my editor, Mark Frauenfelder. Mark has been an inspired teacher to me since we first worked together in 1995.

And then there's the striking visual quality that honors each maker — due entirely to the long hours and (there's no other way to say it) righteousness of designers David Albertson and Sara Huston. Their role is huge in such a book. I want to thank my researchers Anne Moore Odell and polyglot James McQuoid (who would translate Russian in the morning and Japanese after supper). Keith Hammond was an outstanding copyeditor and traffic cop. Shawn Connally used her management finesse to keep the book moving as well. And no one was more crucial than our photo editor Arwen O'Reilly ("A bulldog on the haunch of makers everywhere," as David put it). Have you ever hounded someone to photograph themselves waist-deep in a dumpster while keeping track of the submissions of 90 other people?

I want to thank my wife, Eileen, for sharing her optimism about people as long as I've known her — and for keeping my spirits up while I wrote the book. And I want to thank my parents. My dad opened the wonders of his Snap-on Tools truck to me when I was a kid, and I think it left a lifelong curiosity about what people actually do with all these things.

» Bob Parks

CONTENTS

INSECTS SHOULD NOT READ THIS BOOK

"Man is a tool-using animal. Nowhere do you find him without tools; without tools he is nothing, with tools he is all." When British historian Thomas Carlyle wrote those words in the 19th century, he was telling it like it is. In those days, people were expected to mend their own clothes, fix harness, repair machinery, make wooden toys, and build their dwellings.

Sadly, in these days, most people don't use tools. They don't need to use them, because buying new things or hiring specialists is usually cheaper (both money- and time-wise) than making, modifying, or fixing them yourself. What we've gained in terms of convenience is offset by a growing sense of disconnectedness with the world and a diminished understanding of how things work. It's hard to learn anything by simply consuming someone else's products all the time.

That's why I find this book — which was inspired by MAKE magazine's popular "Made on Earth" section, a kind of "Faces in the Crowd" for amateur tinkerers and inventors — so exciting and reassuring. There actually are people out there — not many, perhaps, but enough to give me hope for humankind — who do use tools, and who use them to create things that are far more useful, ingenious, repairable, and charming than store-bought counterparts. The successful launch of MAKE magazine lends credence to the idea that there are more makers out there than we know.

The makers profiled here are intensely curious about many subjects. Their areas of interest know no boundaries: The guy who built a networked cat door to track his pet's comings and goings also designed and constructed a candle-powered Stirling engine. The woman who hand-weaves wire-and-cloth circuitry into electronic garments also modifies kids' toys to turn them into musical instruments. The fellow who made a coin-operated dog-walking arcade machine is also an accomplished wood carver and furniture maker.

These highly evolved people remind me of a favorite quote, from Robert Heinlein's novel *Time Enough for Love*:

"A human being should be able to change a diaper, plan an invasion, butcher a hog, conn a ship, design a building, write a sonnet, balance accounts, build a wall, set a bone, comfort the dying, take orders, give orders, cooperate, act alone, solve equations, analyze a new problem, pitch manure, program a computer, cook a tasty meal, fight efficiently, die gallantly. Specialization is for insects."

Are you a tool-using animal or an insect?

» Mark Frauenfelder, editor-in-chief, MAKE

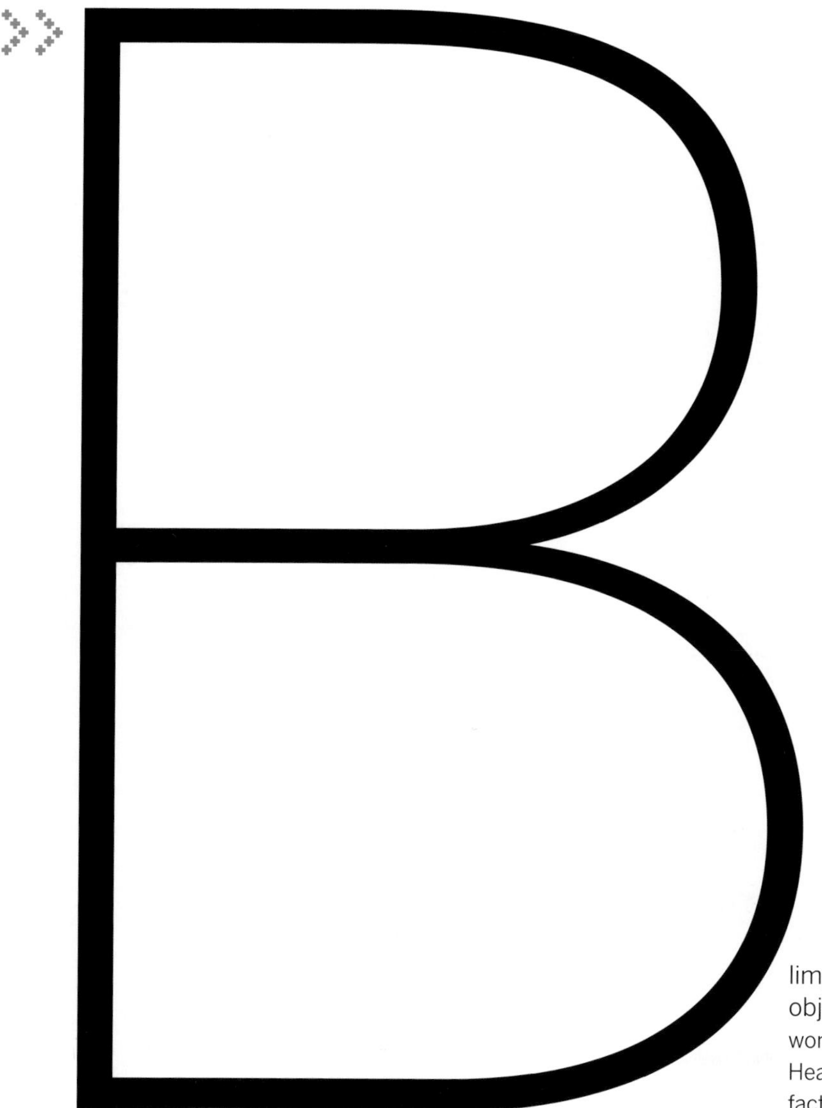

limps are the largest ungrounded objects on Earth — electricity simply won't discharge through them," says Heather Dewey-Hagborg matter-of-factly, as if everyone knows such things.

SKUNK P46

HEATHER DEWEY-HAGBORG P156

JAMES MONACO P172

SARAH MIMS P86

KERRY McLEAN P18

J.P. BROWN P116

KELLY DOBSON P66

MATTY SALLIN P48

LIMOR FRIED P140

MIKEY SKLAR P80

RENÉ GRAU P114

ROCKWELL SCHROCK P34

HIDEO TAKANO P50

TOM JENNINGS P142

ANDY GUSTAFSON P134

SIGURD KIMPEL P166

It's a familiar tone among serious makers of strange electronic devices and mechanical contraptions. Dewey-Hagborg, 23, creates miniature blimps using microcontroller chips and wide swaths of silver Mylar.

This builder of floating robots studied multimedia in college and witnessed big software advances on the web. But one day, she felt the urge to look at technology from the bottom up. In her sleepy college town during summer break, Dewey-Hagborg bought a Forrest Mims electronics book and a RadioShack soldering iron, and built the circuits thousands have practiced before — oscillators, 555 timers, op-amps.

Dewey-Hagborg has been working with electronics for only a few years, but her practical skills and her desire to poke under the hood of high technology mark her as part of a tribe. It's a group that's united by a love of offbeat science and technology; of underdog inventors like Fuller, Farnsworth, and Tesla; and certainly of arcane facts. Some know about the physics of blimps; others about the speed of backyard insects in feet per second. You might hear a welder of custom bicycles in Boston casually mention that, "You know, Napoleon used a lot of aluminum." Or a guy who buries magnetic sensors in his Colorado backyard point out that, "Lightning strikes the globe 100 times per second." It's always phrased in a way that says *of course* you knew that.

A renewed interest in amateur engineering is creating big changes for the culture of technology. A once-small gathering called Dorkbot, for instance, has grown to 20 cities around the world since its launch in 2000. Now thousands of adults attend these informal monthly show-and-tells for electronic and robotic contraptions.

Another indicator: RadioShack appears to be going after the hobby market again after shedding educational kits and publications for years. The company's Vex robotics kits are among its top five strategic moneymakers for Christmas 2005, according to a report by Banc of America Securities. "Electronics has hit a point where normal people can imagine having a hand in making something really amazing," says Dorkbot founder Douglas Repetto. "In the last few years, access to technical information, tools, materials — and peer groups — has really brought a surge in innovative activity."

In many cases, the urge to experiment with simple machines and electronics is a reaction against the sophistication of the PC. According to Larry Lemieux, a longtime electronics observer and publisher of *Nuts & Volts* magazine, people are coming back to electronics after a 20-year romance with computers. Gearheads turned into codeheads, but they eventually returned to electronics when desktop computers themselves became tools to prototype new circuits. "The cycle has run its course," says Lemieux. "Now, with inexpensive programs like SPICE, you can test your designs with an oscilloscope and signal generator on your PC screen."

Other trends encourage experimentation, such as school robotics competitions and televised robot wars. But for many engineering-minded people, the decision to get back to basics was an economic one. "In 2001, a lot of engineers were suddenly jobless — in the Bay Area alone there were 100,000," recalls Tim Sanghera, vice president of Mouser, the second-fastest-growing parts catalog on the web. "The downturn in the marketplace accelerated a period of the independent inventor."

Put in context, the latest wave of backyard and garage tinkering is relatively small in mass-market terms. According to representatives from Mouser and its rival, parts catalog Digi-Key, only a small percentage of customers are students and amateurs. Author Forrest Mims hasn't seen a boost in the sales of his electronics instruction books, although many in the following pages mention them enthusiastically. "Yes, there is a subset of people, but I'm afraid that overall the number is diminishing," says a rather cranky Mims. "RadioShack still sells two of my lab kits. Fifteen years ago, we would sell 100,000 at Christmas. Now we don't sell anywhere near that." It may be that U.S. culture has seen a decline in focus on engineering and science, going back decades. A few tens of thousands of basement engineers are really a blip in the vast market.

Any official personality test for makers would have to include checkboxes for "weekly dumpster diving habit," "pilot's license before age of 18," "recumbent bike," and "amateur radio handle doubles as email username."

The community of makers may not be a sweet spot for business so much as an interesting cultural shift (or tangentially, a useful model for corporate R&D directors). Makers are united by a common set of attitudes, which includes sharing code and parts openly on web forums and homepages. These virtual spaces have replaced amateur radio hamfests as a means to compare notes. (According to online convention, before you actually describe a new project, you must jot down all your components, model numbers, and prices. It's like emptying your pockets of jackknives and brass knuckles before entering a seedy bar.)

Some components are more common than others. More than a third of the 91 makers we interviewed employ a type of electronic chip called an 8-bit microcontroller. In every household, dozens of these low-power chips run everything from cordless phones to washing machines. But in the early 1990s, the chips were off-limits to those outside big electronics manufacturers. You couldn't just use them to create an automatic cat door or a light-up electronic billboard for your car. "If you were lucky, your university would have the $5,000 license to use the C compiler," says Dorkbot's Repetto. "What artist or hacker or geek is going to have that?"

Ease of use finally came from a company named (conveniently enough) Microchip. Its CEO, Steve Sanghi, led the development of inexpensive compiler software that would let anyone program in the familiar language, C. "Microchip set out to create a culture around the 8-bit microcontroller, as strange as that sounds," says Mouser's Sanghera. Sanghi sold adapters that would let you connect your computer's serial cable to the firm's PIC brand chips. Other players, such as Atmel, began to bring out similar products. The companies started giving away free samples as frequently as twice a month to anyone who signed up on their websites.

Sanghi watched it slowly take hold. "I was going to Boston to talk to a company about putting chips in their tennis shoes," he recalls. "Then I saw PICs engineered into little light-up earrings for tongue-piercings. I said to myself, 'What the hell?'"

The change was more abrupt for users like Zach Radding, a robotics teacher in Berkeley, Calif. "All of a sudden, microcontrollers were so much easier to use." He remembers earlier devices with light-sensitive memory; if you wanted to rewrite your code on the chip, you had to expose it to a special UV light to wipe it clean. In 1997, the chips became flash-memory based, meaning that they automatically erased once you loaded new code. "For me it meant I could design, code, and test a project in a just few hours instead of a few days."

Other factors opened unprecedented access to materials. With the rise of online electronics distribution, even exotic parts became as easy to obtain as CDs on Amazon. Companies boasted no minimum order and huge warehouses for next-day shipping. As a female engineer notes: "[Top online seller] DigiKey is more dependable than your boyfriend."

But what really motivates these citizen engineers, once they're given access to the right ingredients? Theories abound as to why people like to experiment with electronics and gear. Many point to the mother of invention — necessity. Certainly, there are examples in the following pages: A programmer in Arizona builds a water circulation system in his barbeque to heat his pool. A doctor in India creates an infant warmer out of scrap wood and light bulbs to lower the mortality rate in her hospital. Recent business books such as Eric von Hippel's *Democratizing Innovation* have argued that users modify products to address deficiencies in performance or features.

But the impetus can't be entirely about solving problems, or makers would cease work once they were satisfied with the results. Instead, many makers continue to enhance their projects beyond any requirement to do so. Serial makers constantly start new projects, across disparate fields and disciplines. Even the doctor from India has moved on from infant warmers to a solar-powered instrument sterilizer.

Garage projects are often derided for being patently "useless" or playfully ignoring the problems they were designed to solve. But that's often the point. Open-ended engineering is a form of play, and its best spokesperson may be Johan Huizinga, the Dutch anthropologist who wrote the 1938 book *Homo Ludens*. Huizinga defines a wide variety of human activity as play, and argues that it should be treated as an important cultural form with no purpose outside itself. Play occurs in a defined space, has rules, and exhibits tension between the start and finish.

Amateur invention certainly fits. Makers follow a desire to challenge themselves with new technologies or materials, and compete with others in their field. On the high school level, science fairs and the FIRST robotics meets have inspired tens of thousands to join student teams. For adults, the spirit of competition is more subtle, but it's clearly present in online forums for nuclear fusors, pulsejets, and personal submarines. Events such as Dorkbot nights, the Burning Man festival, and Seattle Robotics Society meetings also spur makers to produce ever more unusual and astounding machines.

Even solitary makers strive against themselves to follow their own strict rules and heighten the challenge: some Lego builders won't use anything but standard blocks to create sophisticated machines. An engineer in Mountain View, Calif., limits himself to parts found in local dumpsters to make Tesla coils. (Any official personality test for makers would have to include check-boxes for "weekly dumpster diving habit," "pilot's license before age of 18," "recumbent bike," and "amateur radio handle doubles as email username.")

Makers are quick to describe how their long-term projects give them pleasure, but as theorist Huizinga points out, play does not have to be lighthearted and fun. Many builders consciously introduce a high level of risk into their projects for philosophical reasons. Experimental jet engines, high-speed motorized vehicles, and high-energy devices such as nuclear fusors and disruptive discharge coils bring relevancy and vitality to scientific pursuit, claim these makers. "Today's high school science fairs are limited by ridiculous rules," says author Forrest Mims. "Where are the high-powered lasers, the advanced chemistry experiments?"

While some makers aim for a spirit of serious discovery, others deliberately play with culture. They create projects with an anarchic spirit of parody. Matty Sallin's alarm clock, for instance, cooks a piece of bacon to wake you up — it's a joke on our fast-forward consumer culture. (Imagine the press release: "This aroma-augmented clock features an enhanced 'pig' form-factor.")

There is an undeniable satisfaction in getting a chance to press the On button for the first time. For many, it's a moment of confirmation. A signal that an individual can do something that has never been done before, under his or her own steam.

Handmade objects themselves are a rebellion against consumerism, according to comments from several makers. It's a stance that goes back to John Ruskin's Victorian ideas on the value of imperfection and handicraft in Gothic works.

On the following pages, many professional programmers and web developers present life stories in which they describe their garage activities as an antidote to the increased sophistication of high-level languages and multimedia applications. They're finding a newfound appreciation for the real, 3D world. "Physical artifacts appeal to our senses," says Tom Jennings, a programmer and a restorer of classic cars. "There's something deeply satisfying in making things. You have a relationship to your handiwork."

Most of the following projects took weeks or months of hard, physical work. Although this particular form of play isn't well recognized outside the small creative group, there is an undeniable satisfaction in getting a chance to press the On button for the first time. For many, it's a moment of confirmation. A signal that an individual can do something that has never been done before, under his or her own steam. And when a new project is posted to a website, the maker inadvertently launches a dozen other efforts by those who would wish to better the design. After all, there's nothing more contagious than a moment of well-oiled, mechatronic brilliance.

Makers: Location

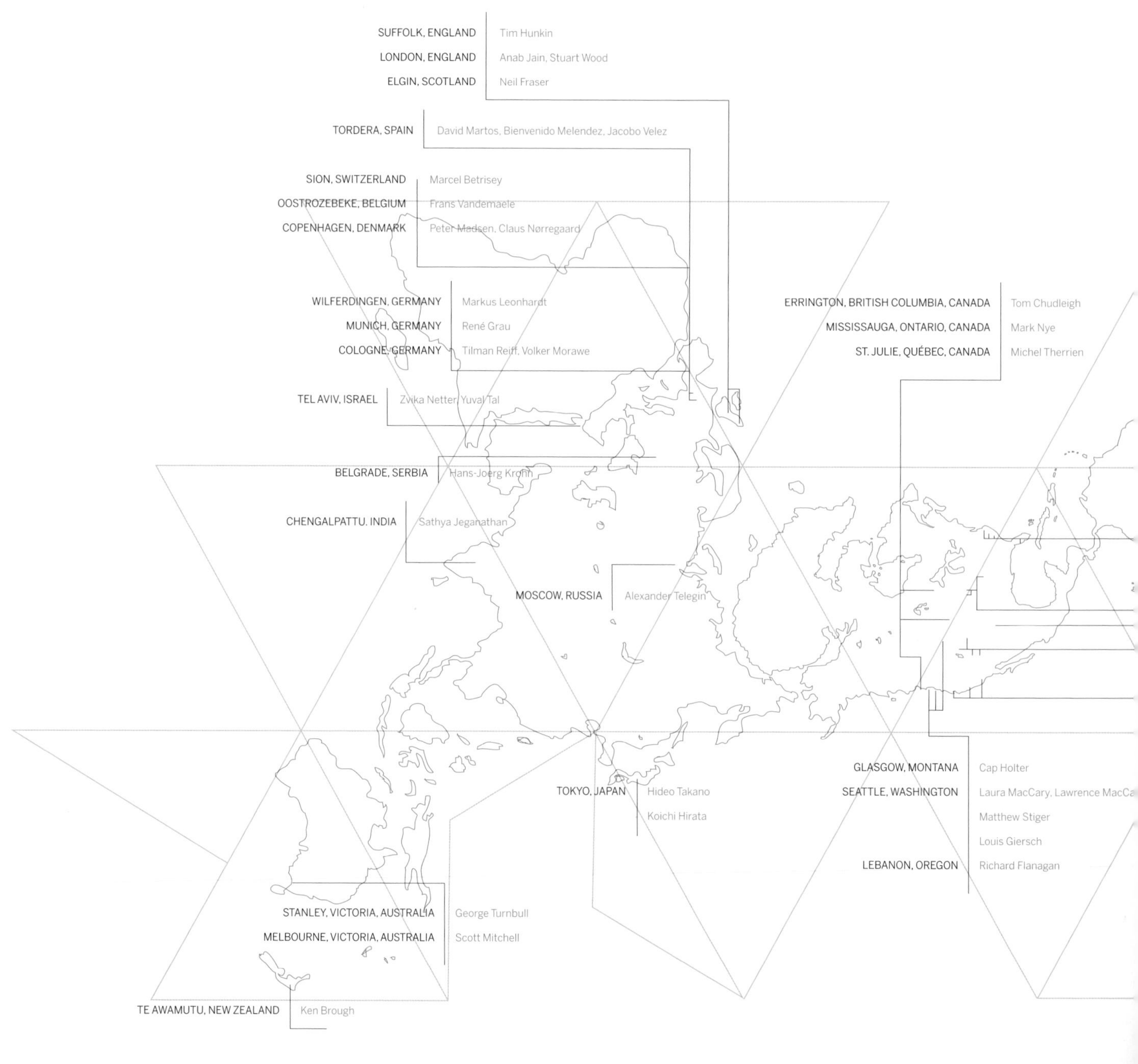

PITTSFIELD, MASSACHUSETTS — Sigurd Kimpel
BOSTON, MASSACHUSETTS — Limor Fried
Skunk
ARLINGTON, MASSACHUSETTS — Ted Selker
CAMBRIDGE, MASSACHUSETTS — Kelly Dobson
EASTHAM, MASSACHUSETTS — Rockwell Schrock
PHOENIX, MARYLAND — David Hess, Jason Bennett
ROCKVILLE, MARYLAND — Owen White
BETHESDA, MARYLAND — Seth Goldstein

RICHMOND, VIRGINIA — Richard Hull
NEW HAVEN, CONNECTICUT — Thiago Teixeira
DUNEDIN, FLORIDA — Ryan O'Horo
COZY LAKE, NEW JERSEY — Al Gori
KINGSTON, PENNSYLVANIA — James Monaco
NEW BERN, NORTH CAROLINA — Larry Cotton
NANUET, NEW YORK — Slava Persion
NEW YORK, NEW YORK — Mikey Sklar
Tristan Perich
Heather Dewey-Hagborg
Matty Sallin
Douglas Repetto

KNOXVILLE, TENNESSEE — Dan Bowen, Mike Coffey
CHICAGO, ILLINOIS — J.P. Brown
MILWAUKEE, WISCONSIN — Erhardt Wowerat
CHEBOYGAN, MICHIGAN — Dennis Havlena
ANN ARBOR, MICHIGAN — Nate Strech, Joel Wollborg, Torrey Baumstark
DAVISON, MICHIGAN — Don Dunklee
WALL LAKE, MICHIGAN — Kerry McLean

FORT COLLINS, COLORADO — Dimitri Merrill
Joseph DiVerdi
SOCORRO, NEW MEXICO — Harald Edens
TUCSON, ARIZONA — David Forbes
Ed Bertschy
MESA, ARIZONA — Todd Harrison, Veronica Harrison

CHICO, CALIFORNIA — Andy Gustafson
SACRAMENTO, CALIFORNIA — Rob Cockerham
SANTA ROSA, CALIFORNIA — Katie Barmazel, Sarah Davis, Ivy White
Bob Blick
SAN FRANCISCO, CALIFORNIA — Jack Buffington
BERKELEY, CALIFORNIA — Zach Radding
MENLO PARK, CALIFORNIA — J.B. Straubel
MOUNTAIN VIEW, CALIFORNIA — Greg Miller
SANTA CRUZ, CALIFORNIA — Bathsheba Grossman
LOS ANGELES, CALIFORNIA — Tom Jennings
WESTLAKE VILLAGE, CALIFORNIA — Paul Taylor
CORONA, CALIFORNIA — Scott Metoyer

AUSTIN, TEXAS — Liz Zazulak, Reagan Ward
DENTON, TEXAS — David Anderson
SEGUIN, TEXAS — Sarah Mims

Best
Melvin's Hardware
RECREATIONAL LEISURE CORP.
Aristo-Cote
panda PRECISION INC
Rocket Roadster
USA
panda

Doughnut R&D

Gasoline-powered monowheel

MAKER	**Kerry McLean**	COST	$10,000
OCCUPATION	Metal fabricator, machinist	TIME	200 hours
LOCATION	Wall Lake, Michigan	MORE	kerrymclean.com

Kerry McLean insists his single-wheeled vehicle drives "like an airplane." Sounds like a proud craftsman's conceit until he finally explains what a monowheel is like to ride. "It's three-dimensional like flying," says the Wall Lake, Mich., metal fabricator and machinist. "You shift your body to steer it. Your pitch changes when you brake or accelerate. You may be hauling ass, but you feel like you're floating."

Only a handful of people have made monowheels in the last century. They're really more for fun, though you can register them with the DMV and buzz your neighborhood, and they apparently do well at parades. McLean has made all kinds of wheels, including a 225-horsepower racing hog (pictured). For $10,000 each, the monomaniac will build a spunky 15-horsepower monowheel for private sale. It's labor intensive; each of the eight wheels he's sold has taken about a month to complete.

To make the giant wheel, McLean starts with a thick aluminum tube bent into a 6-foot series of spirals, like a huge spring. A metal shop does the initial work of bending the tube and is careful not to crimp the inside edges, but the spirals aren't perfectly round at first. In his detached workshop in front of his house, McLean cuts pieces off the aluminum tube until he has a single circle with flush ends, and then he clamps the tube to a giant steel slab. "I gasket this thing down in there, bump it around, heat it, stress relieve it, and weld in place balance lugs. I work on it until I get it flat and round," he says. McLean connects the ends of the circle together by welding an aluminum lug into the center. (To compensate for the lug, he welds a scrap piece of equal weight to the opposite side.)

The rubber tread for this new wheel originates as several new motorcycle tires, which McLean cuts up and sticks to the gargantuan new rim with special primers, adhesives, and more than 200 hand-applied buttoncap screws. He modifies a brand-new Briggs & Stratton 5-horsepower engine to crank out 15 horsepower and suspends it inside the frame by soft urethane mounts connected to steel rollers. The frame and outside wheel then glide along the rollers with a gentle suspension.

Oh, there's one extra step: McLean pours molten lead in empty spaces under the engine. The counterweight prevents gerbilling — monocyclists' argot for accelerating too quickly and sending the driver into a loop-de-loop around the inside of the monowheel.

When McLean is between larger projects, he can't seem to keep his hands off items around the house. A few years ago, he scooped out the innards of his 6-year-old's pink Barbie Jeep, inserted a 5-horsepower motor and chromoly frame, and retroed the body back on (top speed: 35 mph). In his living room, he's modified a vintage Kranich & Bach grand piano with "drag slicks, Corvette header pipes, tachometer gauges, and a '32 grille with a blower." He describes it while playing a short blues shuffle over the telephone.

McLean is charming, albeit a little intense. He built his first monowheel in 1970 and has been obsessively perfecting the design ever since. "I don't feel like anyone has seen it through," he says, affirming a strong dislike for scavenged parts. "They're using some hokey old engine and they think that's good enough. They're limiting themselves. You hear the words 'trial and error.' That's just some hillbilly stuff. Broomsticks and baling wire. I'm doing R&D."

Peter Madsen emerges after the first water test in April 2005 (above). The console below the hatch contains a GPS plotter and two-way joystick for the side rudders. Left, the vessel has just been pulled from Copenhagen Harbor by a crane. It's logged over 50 dives since the first test.

The Life Subaquatic

Personal submarine

MAKER	**Peter Madsen and Claus Nørregaard**	COST	All materials donated
OCCUPATION	Metal fabricator, welder	TIME	3 years
LOCATION	Copenhagen, Denmark	MORE	submarines.dk

Since its launch in May 2005, Peter Madsen, 34, and Claus Nørregaard, 32, have made more than 50 dives in their 41-foot, 6.6-ton, hand-built submarine. It raises a few eyebrows from pleasure boaters and commercial fleet crew, but it gets the most attention from U.S. Navy crew stationed in Copenhagen Harbor. "All the guys come out to look on the bridge wing of their battle cruisers. I think they look disturbed when they see this midget submarine racing past," says Madsen.

Part of the fun of owning your own submarine is doling out the 20,000-league view to friends. The craft can hold up to three people in its 3-foot-wide hull and hits depths of 50 feet. You generally start a cruise at the surface, looking through portholes on the steel tower. To dive, you turn a valve to flood the ballast and trim tanks welded into the hull, thereby reducing buoyancy. Thanks to a small digital camera on the periscope mast, you can steer using a widescreen LCD when submerged or can also peek through the deck hatch above the pilot's seat.

These two freelance Cousteaus toiled for three years in the bay of a shipyard to make the craft. Madsen, a self-employed metal fabricator, started the submarine craze when he launched a smaller sub (25 feet, 4.4 tons). Nørregaard, an independent braze welder for artists, liked the first design, but thought they should collaborate on a new version that would look more posh and feature a tongue-in-cheek nod to a World War II U-boat. The result was their current sub, named Kraka after a wise and resourceful Norse goddess.

The sub's hull came from a huge steel cylinder donated by a company that makes stands for wind turbines. (Almost all of the steel and other materials are donated in exchange for free rides and a plug on the side of the vessel.) They cut two window holes in front to look like torpedo tubes and built a working dive hatch, which allows you to look down into the sea or send a diver out during a cruise.

Non-nuclear subs are like hybrid cars — many have both a diesel engine and an electric motor. Since a combustion engine can't operate without oxygen, it's the diesel that runs above water, turning the propeller and running an electric generator for a bank of batteries. When the craft dives, a mechanical transmission switches the drive to the electric motor. The builders are currently installing an innovative hydraulic thrust system that would make the transmission simpler and obviate wear and tear on metal gears. "In the world of amateur submarines, I believe it represents a first," says Madsen.

Madsen's fairly one-of-a-kind himself. He says a typical home life doesn't really fit with his research goals. He lives in an 18-ton boat in Copenhagen, and says his expenses are low. He makes a living on large metalworking contracts, such as a recent wind tunnel for the Danish Technological Institute. What free time he has goes into the sub. "I live in a very Stone Age kind of way," he says. "I have — I had — a girlfriend. I've had a number, I guess. There tends to be a certain flow."

His life's dream is to live underwater on a semi-permanent basis. He, Nørregaard, and three other experienced maritime engineers are working on an 80-foot, 30-ton submarine, one you can walk around in. "When it's launched, I'm going to move in and sail around in it," he says. "Not just in Denmark, but places all over on the planet."

Space Camp

High-altitude weather balloon with tracking payload

MAKER	**Dan Bowen and Mike Coffey**	COST	$500
OCCUPATION	Macintosh systems administrator, student	TIME	3 months of evenings and weekends
LOCATION	Knoxville, Tennessee	MORE	sunsite.utk.edu/~mcoffey/ux-1/

Amateurs have launched helium-filled weather balloons with electronic payloads before, but never as inexpensively or with as much enthusiasm as Dan Bowen and Mike Coffey. The two young aerospace fans, both 25, wrote code until 4:30 a.m. the night before the launch, and had several tense episodes (they were briefly detained by the security force at a nuclear power plant). Then, after launch, their team of 30 (all members of the local amateur radio club) rushed off in six cars to track down the payload using two-way ham radios. It was an adrenaline-fueled 170-mile chase across Tennessee.

"It felt like a NASA mission to another planet," says Bowen, a Mac sysadmin who works for a nearby city school district. "It's fun to design the payload and bring the parts together, but my favorite part is after launch. It's totally out of your hands and everyone is on the radios constantly comparing positions."

With two successful launches to 50,000 and 60,000 feet, Bowen and Coffey are part of a small hobby community that has tinkered with weather balloons for almost 15 years. You'd think sending a hard metal object into commercial airspace would bother the FAA, but it's perfectly legal. The sheer size of the airspace brings the chances of a mishap to nil. Besides, it's a cool experiment. Bowen and Coffee's first launch, dubbed UX-1, included a GPS receiver and radio to transmit the craft's coordinates and a digital camera to snap pictures. On their website are photos from the flights showing the blue curvature of the Earth (left).

The experiment was cheap, too. "We knew we didn't have the money to do it the way anybody'd done it before," says Coffey, a marketing student who has a hint of a Southern accent and the smooth conversational style of a NASA PR director. First, they bought weather balloons online ($30 each, rated to burst at 50,000 feet), and a parachute that would bring the payload down safely.

At RadioShack, they picked up a closeout GPS, the DigiTraveler, for $20, which they opened up and soldered to their radio transmitter. The transmitter itself came as a kit bought on eBay that Coffey assembled into an Altoids can and tuned to 144.39MHz, a standard frequency for GPS tracking. In fact, the frequency is so common that they consulted a few free, online 144.39 tracking servers to watch their balloon during recovery. The flight computer was a Basic Stamp II microcontroller, available at any robotics supply house, and it was responsible for controlling the GPS, starting a strobe light and siren as it was falling to the ground, and firing the camera, a $10 Dakota Digital.

The morning they launched, Bowen knew there'd be trouble. The expensive lithium batteries in the payload completely died. He quickly soldered together five 9-volt battery connectors, substituting heavier alkaline cells in the circuit. But with the extra weight, the balloon wouldn't ascend as fast, and the GPS controller was only programmed to send location data for two hours. A slower ascent meant that the device would likely be lost.

Pressing their luck, the crew decided to launch anyway. After three hours, nine chase cars continued to circle various projected landing sites. "Forty minutes had gone by since the last data packet had been heard from the balloon in flight, and the mood in our car was pretty glum," recalls Coffey.

Then one ham radio club member received a tracking signal. "A radio yelp vibrated off the car's dashboard!" »

NIKE

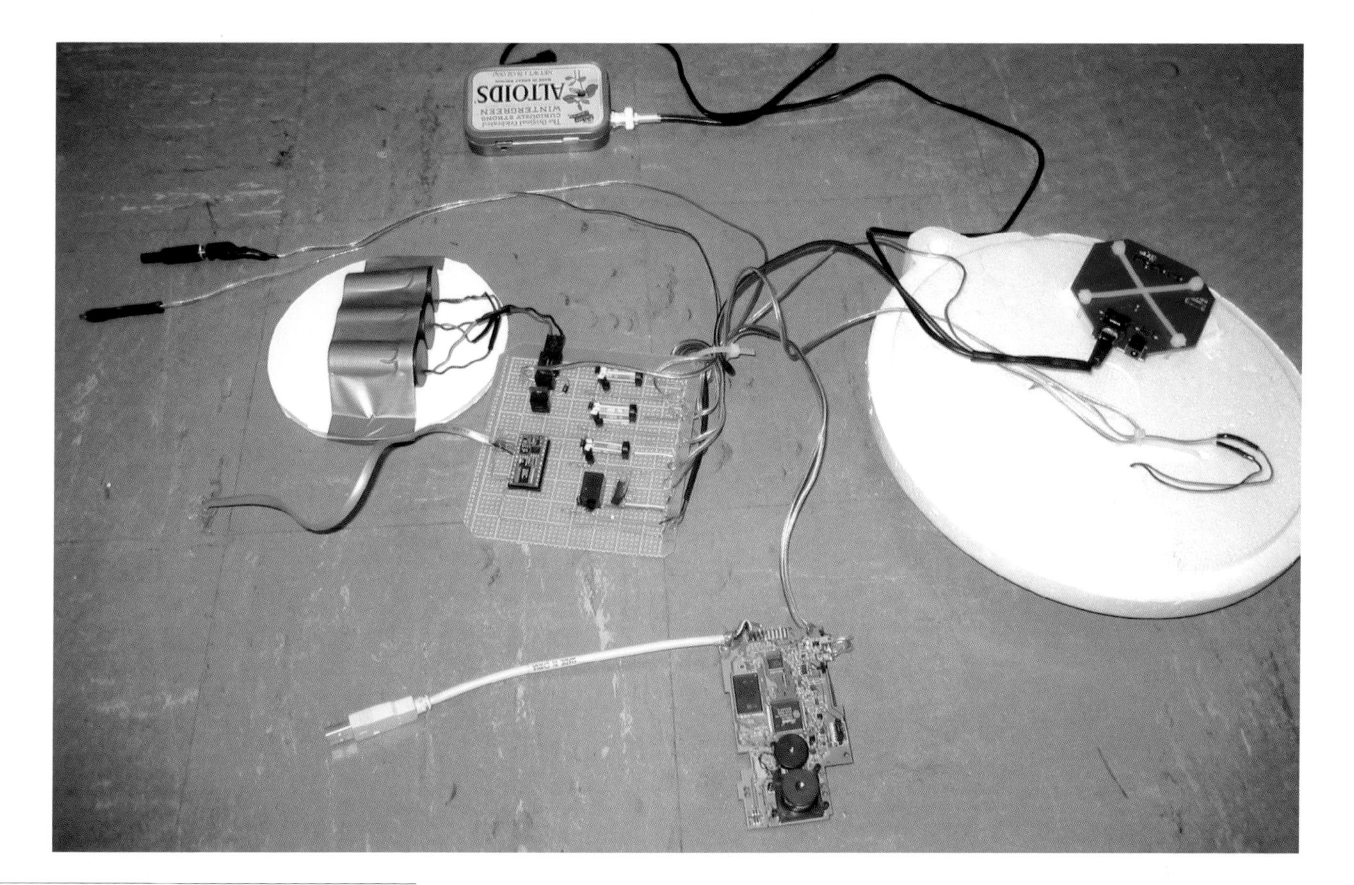

"It went up like a spring out of my hand," says amateur balloonist Mike Coffey, 25. The two-pound payload ascended quickly at 900 feet per minute, reaching a height of 60,000 feet. That's about half the altitude of the top amateur record holders, but a respectible height for the team's second try. This time, they used a hockey-puck-shaped $75 Garmin 18 GPS (above) and a radio transmitter stuffed into an Altoids box. The balloon (left) flew for two hours, automatically shooting brilliant images of the Earth's curvature, and finally landing via parachute in Sweetwater, Tenn.

» recalls Bowen. The payload's onboard computer had mysteriously rebooted during flight, restarting the software routine and sending final tracking signals. The coordinates put the balloon somewhere in a Rutledge, Tenn., field. The weary crew asked neighbors for permission to enter, and soon heard an audible alarm chirping in the rippling grass. "UX-1 had cheated fate!" says Bowen.

After their recovery ordeal, Bowen and Coffey went directly back to their houses and to bed. But there was one thing that Bowen did before catching up on sleep — uploaded the dozens of photos of space onto the radio club website. He says: "You just have to hook that camera to the computer and see whatcha got."

Betrisey adjusts his clock with the vacuum chamber removed. The 33-pound glass bell sits to the clock's left and was the most expensive part at $1,000. Its 0.4-inch walls keep a near-perfect vacuum inside so that only light rays push the pendulum.

The Power of Light

Radiometric clock

MAKER	**Marcel Betrisey**	COST	€2,500 ($3,000)
OCCUPATION	Contract electronics developer	TIME	6–12 months
LOCATION	Sion, Switzerland	MORE	betrisey.ch

He may be Swiss, but Marcel Betrisey, 44, wasn't exactly groomed from birth to be a clockmaker. The son of grape farmers, Betrisey spent half his life wandering, backpacking through 67 countries before turning 30. Returning to his small hometown nestled in the Alps, he started a household appliance repair shop and started tinkering with gadgets on the side. He fashioned a CD player from an antique typewriter (keys controlled the functions), and created a pendulum that responded to the Earth's rotational momentum. In 1997, he turned to clocks and discovered his medium of choice.

To prove himself to the clockmaking world, he welded and machined a 660-pound timekeeper driven by 95 large ball bearings. Then, he made one from stainless steel pipes and an old saucepan; tiny gusts of air kept the pendulum in motion. Lacking a clockmaker's tools, he created his own, including saws hacked together from a photocopier motor."I don't know why I always need to improve every tool I have, even the new ones. It must be a kind of disease, I guess," he says.

One day in 2001, Betrisey decided to take on the biggest horological challenge of his career. It would be a radiometric clock, using a little-understood trick of visible light to keep the pendulum in motion. In the 19th century, scientists discovered that if you shine a bright light on a thin black element inside a vacuum, the power of light will give it a little push. Though some scientists disagree on why this works, most accept the theory of "thermal creep." Inside the vacuum chamber, small volumes of gas heat up and slip around the edges of the thin sheet, creating unequal pressure.

"It arose out of an unsolved question," says the clockmaker. "That was the most interesting." Betrisey knew he needed to keep the pendulum swinging with precision, so he attached a bar of ivar, an alloy prized by clockmakers for its resistance to distortion in various temperatures. He drilled and welded stainless steel globes to the pendulum's bottom for weight, and glued on four tiny mica sheets. Betrisey then blackened the mica with candle smoke, sealed up the glass hemisphere with silicon glue, and brought the pressure inside the clock to 0.01 bar with a vacuum pump.

In his repair shop, Betrisey started the clock with a powerful magnet. The radiometric effect applies a tiny force to keep the 4-kilo pendulum moving. Once swinging, two 35-watt halogen bulbs mounted to one side flashed when the pendulum swung away from it. When the pendulum swung toward the lamps, an infrared sensor triggered a relay to turn on halogens on the opposite site, giving it a push in reverse.

"It started!" Betrisey wrote in his notes of that day. "And it works so well that I do not even dare touch it."

Someday, Betrisey wants to lower the pressure inside his clock a great deal to see if it will perform an entirely different light-powered trick. "I don't have the expensive vacuum pumps to do it yet," he says. "But when the vacuum inside is quasi perfect, the power of photons will move the pendulum — the same power used by the solar sails moving an interplanetary probe."

Until then, repair and engineering jobs beckon. And in December 2004 he had his first child, a little boy who keeps him and his wife busy. The baby's birth started him thinking about his own life wandering and inventing. Will the boy leave school and become a tinkerer like his dad? "Maybe he won't like tools?" Betrisey wonders. "Maybe he'll be a poet?"

Cool Worlds

3D sculptures digitally printed in metal

MAKER	**Bathsheba Grossman**	COST	$250
OCCUPATION	Sculptor, designer	TIME	Up to 5 weeks
LOCATION	Santa Cruz, California	MORE	bathsheba.com

When Bathsheba Grossman, 39, was a senior in college, a professor showed her a whole new type of art called mathematical sculpture. "It blew my mind," she recalls. "I never knew there was such a thing. I walked into his studio a math major and walked out an artist." She wound up completing her bachelor's degree in mathematics anyway — finishing up *summa cum laude* at Yale — and changed her career track to art after graduating. Little did she know she'd have to wait a decade before she could practice mathematical sculpture herself.

Unlike wielding a paintbrush or slinging modeling clay, these complex surfaces were difficult to create with traditional tools. The shapes twist around in undercuts that continue into the center of the object. A traditional bronze mold wouldn't produce the complex geometries. A few years later, after picking up a master's in fine art from the University of Pennsylvania, Grossman worked hard to create similar intricate pieces by sculpting directly in wax and molding in bronze. "I was trying to do the work by hand and wasn't having a good time," she says.

Then in 1997, an inexpensive 3D printing process became available. The technology is similar to inkjet printers, but these machines churn out physical objects based on computer files. The first affordable 3D printing created objects in starch, and Grossman was among the first to use it. "Advanced prototyping went from something that was completely in-house at Boeing to something you walk in off the street and order," she says. Then in 2000, the technology improved again. Starch was interesting, but solid metal would be even better. A new company called ProMetal offered a service that would print objects directly into bronze and steel. It was the culmination of the artist's vision.

Now, Grossman's objects are created on computer, printed at ProMetal, then finished back at her studio. Some of her pieces are simply models of famous mathematical surfaces such as gyroids or Menger sponges. These can be generated on a computer using established algorithms. Other Grossman pieces appear mathematical in their symmetries, but come completely from the artist's imagination and often borrow forms from nature. One of her designs, for example, resembles Romanesco broccoli. Her *Quintrino* features the shapes of the starfish that live a few miles from her woodsy one-acre home in Santa Cruz. (She lives with her husband, an astronomer for the University of California and for NASA.)

In the front yard of their house, Grossman sketches out her pieces in a 10-by-12-foot shed (another identical shed is full of finishing tools for the objects). She'll do several studies using paper or popsicle sticks, then try to rebuild the form in a CAD software app called Rhinoceros. To help with the arduous task of generating intricate surfaces on the metal, she'll write her own computer scripts in Perl. After art school, she worked as a computer programmer for the financial industry, and she says the coding is part of the fun. "I love putting together a couple of mathematical research tools with freeware and 3D modeling products," she says.

Once she sends the digital files to ProMetal, the company prints out her sculptures in layer upon layer of fine steel powder, held in place by a laser-activated glue. When the object is heated, the steel powder fuses together to form a porous version of the complete object. »

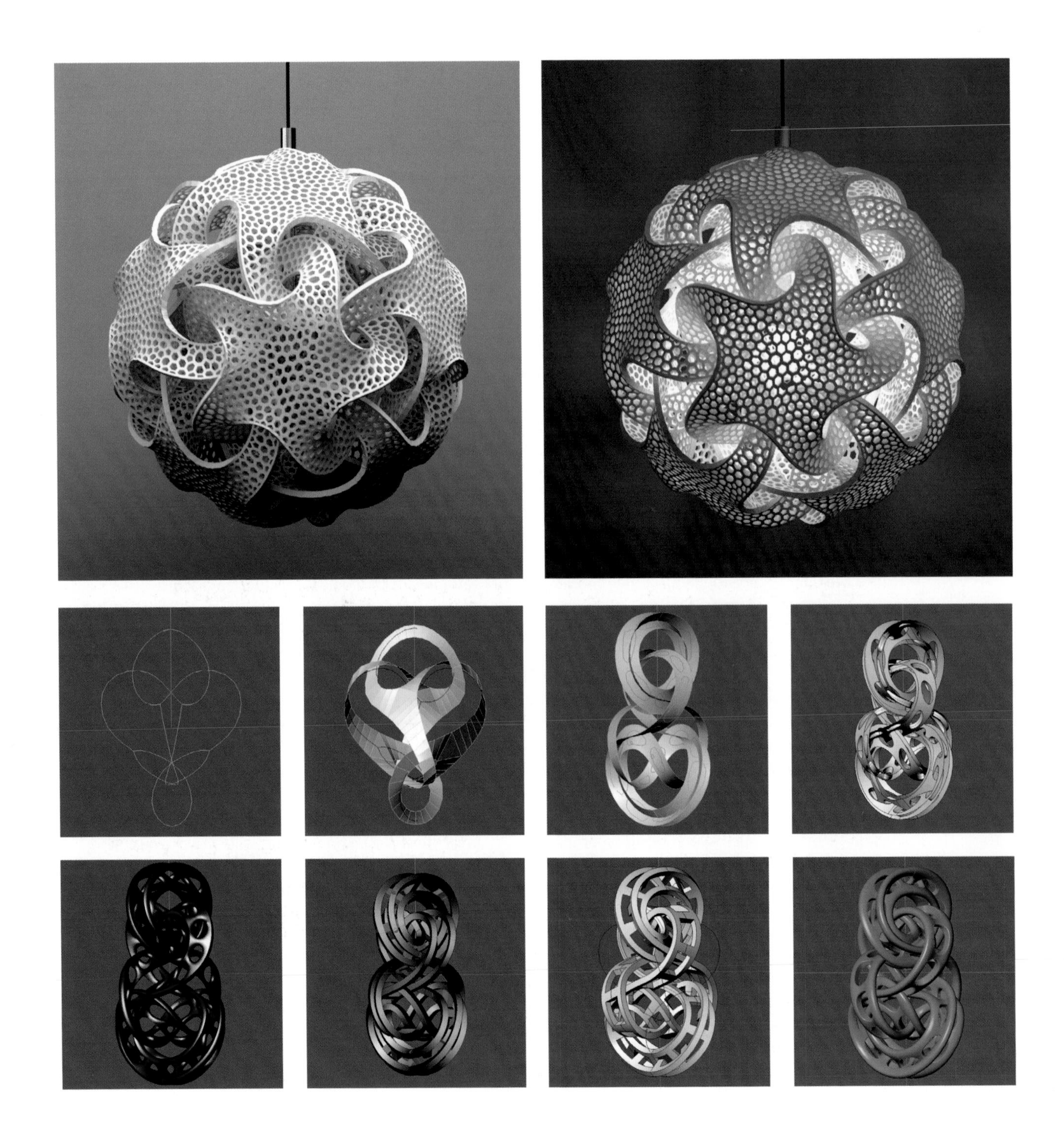

Instant fabrication equipment printed out the artist's 2004 sculpture *Lazy Eight* (below) directly in bronze. Opposite, bottom rows: The original computer file used to generate the artwork. Opposite, top row: Renderings for a new lampshade.

» ProMetal technicians heat the sculpture again and dip its edges in molten bronze. Amazingly, the bronze soaks through the entire object like water through a sponge.

When the artworks come back to Grossman, she still has a lot of work to do. In her shed, she uses an engraving tool and Dremel to hide the manufacturing scars on each piece, then dips the pieces in a pot full of ammonium sulfides and other chemicals to create a brownish patina. Her workshop has an abrasive tumbler to polish the surfaces. Some of the objects go to museums, while others go on sale on her website (she currently makes a living through online sales).

Because of Grossman's bulk orders to ProMetal, it costs her less than $250 to print out each piece. But she notes that anyone can request individual orders for $500, which in her mind is a bargain. "I can't tell you how cool it is to have your own small hunk of metal," she says.

While the high-tech printing process still appeals to Grossman, she's lately branched out into other materials. She's part of a 2006 team at the International Snow Sculpture Championships in Breckenridge, Colo., planning to make a 12-foot version of her *Gyroid*. And she's recently licensed her work to a company that makes high-design plastic lampshades. She simply likes these math-inspired shapes, in any medium. "I feel calm and hopeful in making them," says Grossman. "They give me a feeling of order in the universe."

DURACELL
DURACELL

Tabby Tracker

Networked cat doors

MAKER	**Bob Blick**	COST	Under $50
OCCUPATION	Laser engineer	TIME	A few evenings
LOCATION	Santa Rosa, California	MORE	bobblick.com

When the Blicks' much-loved short-haired tabby, Yum-Yum, went into seizures one weekday morning, the household descended into chaos and worry. A phone call to the vet produced only more questions: Had the cat gone out the night before? Could she have been bitten by another animal?

Luckily, it was Bob Blick's hobby project that provided definitive answers. Blick had networked and motorized the cat doors in his house. He simply went to a nearby Linux workstation to discover that Yum-Yum had been indoors the whole previous night. "There's a lot of panic when there's a medical emergency going on. I had no idea the project would serve a real purpose," says Blick, an engineer who works by himself for an industrial laser manufacturer and who describes himself as "definitely middle-aged." The cat's fine now and the family thinks she may have been exposed to some weed killer days before the event.

Blick first installed the feline entrances at the front and back of the house when skunks and other cats began coming in to eat Yum-Yum's kibble. Now when his cat wants in, she knows to pull on a small colored patch of wooly Velcro on each door (it took ten days of coaxing to teach the cat this trick). The small change in door position triggers a pressure-sensitive switch to signal a PIC microcontroller. The circuit powers up servomotors that swing the door open, and after a few seconds, the microcontroller powers the door closed to the point at which the pressure-sensitive switch signals it's shut.

Did Blick link his doors with a network cable such as, um, Cat-5? No — he only spliced an unused pair in the POTS wire already inside the walls. The PIC sends a signal down ordinary telephone wire to the homeowner's Linux PC in an upstairs bedroom. "I might incorporate that in with the graph, so I could track the cat events," says Blick, who speaks in a sort of deadpan George Carlin voice. "Really though, plotting statistics about cats has to be a really stupid thing to do."

By crunching the data, Blick has learned that Yum-Yum enters the house an average of six times per night and vastly prefers the back door (likely because it is the older of the two installed).

The projects — dubbed "work avoidance systems" for their builder — nevertheless keep him busy. Blick is one of the web's best-known tinkerers. In 1997, he was the first to make a propeller clock, a project which has become something of a rite of passage for many weekend potchkies. When not in motion, the clock is simply a wooden arm studded with LEDs. With the arm spinning in a dark room, the viewer sees a digital readout hanging in the air. The devices had been commercial novelties, but Blick was the first to create one as an amateur.

In some ways, the cat door was a tougher project because it had to blend into the house, and it required advanced animal training. Blick's wife, Carol, didn't care for its looks on their new house. "She said, 'That's kind of ugly. Do you think you could put a little brass doorknob on it? Could it have a window with little panes?'" recalls the dutiful husband. Blick has a woodshop with tools he bought on sale, and wants to learn to use them better. He created the cat window, panes and all. No doubt visitors to his site will soon have networked cat doors with little panes too.

Schrock's diagram for his modified traffic light. The 18-year-old rewired a real streetlight to flash according to the weather or friends' messages on his computer. It currently hangs in his bedroom.

Intersection

Old traffic light flashes weather and message alerts

MAKER	**Rockwell Schrock**	COST	$45
OCCUPATION	Student	TIME	15 hours
LOCATION	Eastham, Massachusetts	MORE	hackalight.blogspot.com

It's a testament to Kathy Schrock's cool-mom factor that she quietly looked on as her 18-year-old son connected a hacked municipal traffic light directly to household current. "She was mainly worried about me frying one of the computers in the house," recalls Rockwell Schrock, her son. His mother picked up the light at a local flea market and presented it to him as a birthday gift. Bonus points.

The young inventor has since programmed the vintage traffic signal to flash whenever a thunderstorm or hurricane is due to hit his small hometown on Cape Cod. It displays barometric pressure in three stages, and flashes at him when friends want his attention in an online chat room. And it's controlled using a parallel cable and code written by Schrock, whose middle name is Buckminster, after Buckminster Fuller, the 20th-century visionary and inventor of the geodesic dome.

Schrock, who often goes by "Bucky," comes from a family of achievers that reads like a well-adjusted version of *The Royal Tenenbaums*. When you call their house, you navigate through a commercial phone tree to reach the various members ("Dial 3 for Rockwell"). He says they set it up because it cuts down on telemarketing calls, but it's a testament to how busy they are. His father, Jim Schrock, is a civil engineer. Mom directs technology for the local school district, and is a well-known author and lecturer on the topic of tech education. And yes, they live in a giant, wood-shingled geodesic dome.

Schrock himself already has several impressive entries in his CV. He once hooked up his Lego Mindstorms set to the family toaster, so anyone could start the device through an online interface. ("I took it down because I was worried people would make toast from China," he deadpans.) On his summer vacation, he runs the network of a local real estate agent and mans the workbench of the repair shop Cape Cod Computer. He plays violin, runs 110-yard high hurdles on the track team, and rides a skateboard (albeit rather awkwardly).

Last fall, Schrock took an independent study in C# (pronounced "C sharp") computer programming at Nauset Regional High School, where he is a high-honors student. During the class he created an online database in which teachers can post homework assignments for the school of 1,100 students. Called "Homework Online," the principal made the site mandatory for teachers to use in the coming semester.

"I really love C#," says Schrock, explaining the basis of the traffic light hack. For weather applications, a script extracts the barometric pressure and storm data from the Yahoo Weather page, directing the appropriate signal through a parallel port. For chat rooms, Schrock's code monitors discussions, looking for keywords that he has shared with his school friends. If someone sends him a message with "Green Flash" in the text, the corresponding light will comply. "It's neat to get data where you don't have to go looking for it," he says of the result. "It's just there. It opens the doors for alternative understanding of the information."

Outside the Box

Spherical wooden treehouse

MAKER	**Tom Chudleigh**	COST	Can$15,000 (US$12,700)
OCCUPATION	Technical writer	TIME	2 years
LOCATION	Errington, British Columbia	MORE	freespiritspheres.com

Living in the trees is not exactly roughing it. Tom Chudleigh, a freelance technical writer, spends many of his summer workdays in a home-built treehouse. The 10.5-foot sphere features a sink, refrigerator, microwave, futon, electric heater, and sound system. (There's no bathroom, so lay off the coffee or you'll be racing up and down the handmade wood-slab spiral stairs all afternoon.)

The current abode, located in the Vancouver area, is Chudleigh's second try at a round house, and it was finished in May 2005. It hangs 14 feet off the forest floor by three ⅝-inch Polysteel synthetic ropes attached to the exterior, and it sleeps four. When the trees move, it gently rocks in the wind. "Sleeping in the forest canopy is awesome," says the builder, 53. He and partner Rosey spend a few days a month living there during warm weather. "We often sleep with the door open just to wake up with the birds," he says.

Chudleigh has held several jobs as an industrial arts teacher and a boat builder. For a time, he lived on a 45-foot steel-hulled ketch, which he built himself. ("Like most boats, it vanished from my life when the marriage ended," he chuckles. "So it goes.") The sphere's interior resembles nothing so much as the galley of a high-end cabin cruiser, with its tightly organized appliances and teak interior. The construction process draws heavily from boat design as well.

The structure comes together like a "cedar-strip plank canoe," says Chudleigh. First, he creates a wooden frame out of Sitka spruce, connecting the pieces until it looks like the skeleton of a globe. He laminates the sphere with two layers of spruce strips running horizontally then vertically, and covers the outside with fine-weave fiberglass cloth, so that the wood finish is visible. When the tough exterior surface dries, Chudleigh borrows a fancy Bosch jigsaw from a friend to delicately cut out the spaces for the five round windows and door. To get the window itself to match the curve of the exterior, Chudleigh creates a round steel form and passes it off to a local skylight manufacturer. The workers heat an acrylic sheet in an oven and then clamp it into a mold and apply air pressure underneath to form a bubble.

All the hardware on the windows and doors is handmade. (Chudleigh made his own garage bronze foundry 23 years ago after studying plans in a book.) His last steps in making the sphere include insulating the interior with 2-inch fiberglass, finishing with upholstery, and adding plumbing and 20-amp, 240VAC service. The treehouse keeps warm in the winter with a small electric heater.

To get the sphere to the woods, Chudleigh loads it onto a custom-welded trailer, and hauls it behind his pickup. For his latest treehouse, he scouted out a good spot in a patch of old-growth forest, and rented the land from a friend. The site is a short two-mile bike ride from the apartment he shares with his partner.

To secure the ropes to the three anchor trees, Chudleigh dons mountain climbing gear (he avoids spurs so he won't damage the trees) and creates loops a few yards higher than the desired hanging level. (He claims the treehouses can be lifted as high as 120 feet in the air this way.) A loop of slightly thinner rope holds the three main lines to each tree in case of an emergency. Chudleigh figures that if a tree falls on top of the house, one of the three ropes will break away and the sphere will swing free without falling. Cable attachments on »

The treehouse sleeps four people in three beds (including a loft above the kitchen sink). The 44-inch diameter picture window, shown here, is made from curved acrylic and opens out a few inches to catch forest canopy breezes. "The acoustics are great inside," says the builder, who likes it as a spot to watch DVD movies.

A few miles up the road in his brother's workshop, Chudleigh finishes applying thin strips of spruce to create a tough exterior shell. Once the windows are chopped out and the cabinetry and upholstery installed, the sphere will weigh 500 pounds.

» the sphere are made with aluminum backing strips so they don't pull through.

Once the sphere is secure, Chudleigh builds a set of spiral stairs and a short suspension bridge. He cuts 3-inch slabs of cedar logs he finds on the forest floor. The support cables go through a spreader bar that pushes against the tree (an old sailboat mast trick), and then down to the outer edge of the stair.

During the day, the sphere makes a great home office to bang out a few training manuals. Most of Chudleigh's writing involves instructional materials for pulp mills, oil refineries, and power generation plants. He studied biology in college, but found job prospects slim, so he went to technical school. Now, he writes and edits in the sphere, and emails his work from home. (The treehouse does not have broadband or a telephone yet.)

With his method down pat, the builder would like sell a few tree spheres to researchers, hunters, or campers. Two other models are currently hanging on other patches of rented forest, including a prototype he started in 1995. His website advertises fiberglass models and DIY sphere kits, but until sales take off, his current job suits his lifestyle just fine. "I like short-term contracts, and then I go out and play treehouse guy until all the money's gone."

As coins revolve on an old record player (top), they reflect light back to optical sensors and trigger various instrumental samples. The music player's circuit (left) consists of four infrared sensors and four infrared LEDs.

The Sound of Money

Music sequencer triggered by coins

MAKER	**Thiago Teixeira**	COST	$50 (plus coins)
OCCUPATION	Student	TIME	A few hours
LOCATION	New Haven, Connecticut	MORE	varal.org/media/coinsampler/index.html

An old record player, $2 worth of RadioShack sensors, and a pocket full of spare change. That's all Thiago Teixeira needed to take a break from a brain-crushing master's thesis at Johns Hopkins University. As the record player spins around, the coins trip infrared sensors in the tone arm and trigger samples on a computer. At different places on the diameter of the record platter, you might get a snare drum or the sound of a horn. Throw on a few at random, and you get a noisy but regular beat. Adjust the coins at better intervals, and you can cook up a pretty good salsa.

Teixeira hails from Rio de Janeiro, Brazil. He's a 25-year-old electrical engineer entering Yale to get his doctorate, working with ad-hoc networked sensors. It's next-big-thing science that looks at devices as small as a piece of rice, each able to receive data and relay it to the next device along the line. The technology could be used for search and rescue, surveillance, or the monitoring of industrial machines. One thing's for sure: it's hard.

The coin sequencer, he insists, was very easy. He soldered two resistors, a transistor, a phototransistor, and an LED onto a small board. The diode shines light onto the black surface of the record platter. When a coin goes by, it reflects light back to the phototransistor and sends a signal to the computer via a parallel cable. Teixeira programmed a small application in C# that creates software in DirectX. Four sensors on the player's tone arm correspond to four different samples.

His inspiration is simple: "There was this movie once, I don't even know what it was. There was this instrument made of rocks cut so they're like cylinders. So you stack them together and now all that rotates, a big record player. It was some movie about some medieval guy, supposedly a magician and he's also a doctor... well, some people think he's a witch. I don't even remember much about it."

The sequencer is clearly a work in progress, depending on how much time Teixeira has and how much he needs to procrastinate on his final paper. He'd like to add a sound recorder to the player so that it would pick up audio from a live performance and play it back via the coins.

Over the last year at school, he and a friend have performed at parties where the friend plays guitar and he plays the a sequencer on a Mac. He'd like to create a housing for the sequencer to secretly use it onstage. That way, he could be the magician.

Scale Model

Robotic fish

MAKER	**Koichi Hirata**	COST	¥35,000 ($300)
OCCUPATION	Research scientist	TIME	2 weeks
LOCATION	Tokyo, Japan	MORE	www.nmri.go.jp/eng/khirata/fish/index_e.html

How does a salmon hit 28 mph with such efficiency? Koichi Hirata, a maritime researcher in Tokyo, knows. He's studied fish for the last 15 years and made more than a dozen working models of them as a hobby. He says salmon generate ripples of energy along their bodies and tails. Muscles in the tail push water to one side, while the tail fin pushes to the other, creating pressure gradients along an S-curve that propel the fish forward.

Despite their simplicity, Hirata's hobby fish can create some of the same pressure gradients as the salmon. His PPF-09 project contains an articulated caudal peduncle — that rear end that helps deliver efficiency. Via a standard modeler's remote control, he can change the frequency or amplitude of the tail movements, make the fish dive, and make it turn. An Atmel microcontroller interprets the transmitter's analog signals for the motors. Unfortunately, when Hirata first tried the robot out, it swam at a mere 0.9 miles per hour (not quite running with the salmon). "This model fish robot is for amusement," he writes in an email in English. "It was too dull, and does not have quick motion like a fish."

Hirata, 38, is no born naturalist. "I was not interested in the sea, and did not grow up near the ocean," he says. "My interest was in thermal engines for automobiles." He became obsessed with the Stirling engine, a design often used in aerospace that moves a piston with heat applied to the engine's outside surface. He applied to his current job as a chief researcher at the National Maritime Research Institute when he heard that scientists there were looking at new Stirling designs. Over the years, Hirata has machined more than 40 different engines from aluminum, stainless steel, and brass. His home — which he shares with his wife and four daughters (2, 4, 6, and 8) — has no space for a workshop, so he generates CAD drawings on the PC and brings the plans to work to carry out.

Hirata began to make robotic fish outside of work, and began to teach metalworking and mechanical design at various Tokyo universities. With his students, he has created dozens of "simple fish-type boats" — which are more like a surface vessel with a fish's tail. The projects all use microcontroller chips to perform simple routines such as speeding up and turning. Hirata first heard of microcontroller-based projects long ago, but wasn't immediately interested. "My father was an electrical engineer," he says. "I touched a computer and some electric equipment when I was 5 or 6 years old. I did not like them. I chose mechanical engineering to avoid the field.

"In 1999, I was 32 years old, and I started to experience the microcontroller again," Hirata recalls. "I was immediately absorbed." In October 2005, he published a book whose title roughly translates as *Introduction to Electronics Handicraft: Robot Fish.* The book is full of encouragement for students and detailed instructions on how to program Atmel microcontrollers, or *maikon* in Japanese. In the preface, he writes: "The fun of 'Making' is about bringing individual ideas to life." Little do his readers know how many lifelike fish he has taught to swim.

A DIY night-vision scope (top) made from an image intensifier (eBay, $27) and dumpster parts. Its images (left) are fun for experiments indoors, but range is limited outside.

Infra Rad

Night-vision scope from discarded parts

MAKER	**Greg Miller**	COST	$39
OCCUPATION	Engineer	TIME	4 hours
LOCATION	Mountain View, California	MORE	sixmhz.com

On a stinking-hot day somewhere in the posh neighborhoods of Mountain View, Calif., Greg Miller, 27, stands waist deep in a dumpster, hunting for his next big project. In the past, he's found a stun gun, dentures, and an Atari game system. He's used a couple of stovetop grease catchers to make a 250,000-volt Tesla coil. A castoff exercise bike became an electricity generator for the shop lights above his workbench. A couple of microwave ovens became a shiny new railgun. Miller, an engineer at energy startup Nanosolar, says the serendipity of garbage surfing helps him determine what great project may be next.

Miller's first dive occurred while he worked at a center for chemical abuse in Iowa, where a local homeless man showed him the art of surveying dumpsters. Since then, Miller, who has an undergraduate chemical engineering degree from University of Iowa, has found that the practice introduces him to new branches of science. Of course, there's the occasional brush with kitty litter, baby diapers, and live raccoons, but he says the Bay Area is rich with trashed technology. "When you get a lot of stuff that doesn't cost you anything, you can fool around with it," says Miller. "I credit my electrical and design sense to the art of dumpster diving."

One recent afternoon, Miller found an 8mm camera lens and a defunct computer monitor. Moments later, the notion of an improvised night-vision scope began to form in his mind. He went home and bought an image intensifier tube on eBay for $27. The component, from 1971, absorbs infrared light and emits electrons, but needs 3,000 volts of alternating current to operate. Miller stripped the flyback transformer from the old computer CRT monitor and connected four D batteries. The camera lens became the eyepiece, and four infrared LEDs he found became illuminators for the front of the scope. The result? Not quite military special ops, but good enough to peer around his house in the dark.

Some of the best pickings are on the Stanford University campus, where his wife, Gillian Zaharias, just finished her Ph.D. in chemical engineering. Although Zaharias won't climb in a dumpster or accompany her husband during the day for fear of being recognized by her professors, she's happy to poke through trash at night with a long stick while Miller fetches items with a Princeton Tec LED headlamp. Since he moved with her from Illinois in 2001, Miller has led 15 to 20 people a month through the local dumpsters. It's akin to an academic seminar on trash, complete with tips and people-handling skills.

"I was standing in a dumpster eating some Pringles, and a lady came to throw out her garbage," recalls Miller. "She couldn't figure out what was going on, so I called into the pile like I'd lost my little brother: 'Hey, Tommy, where are you? Are you down there?'"

The bike builder known simply as Skunk rides his 140-pound modified Western Flyer (left). He and friends cruise around Boston listening to music from a built-in MP3 player and 30-watt speakers. An onboard radio transmitter (below) beams the music to fellow riders nearby.

DJ Skunk

Modified bicycle with FM radio transmitter

MAKER	**Skunk**	COST	$900
OCCUPATION	Bicycle manufacturing supervisor	TIME	Constantly evolving
LOCATION	Boston, Massachusetts	MORE	scul.org

It's the perfect machine to lead a bicycle club through the streets of Boston. First, it's a rolling radio station, broadcasting MP3 songs to radios on other bikes up to 60 feet away. And then the chopped and extended escort vehicle has two 30-watt car stereo speakers bolted to old steel cooking pots, 40-watt halogen lights up front, and a blaring car horn from an old Ford truck. All weighing in at a thigh-busting 140 pounds.

The maker's name is Skunk (yup, that's what shows up on Caller ID when he phones you). By day, he's a seasoned technician of high-end titanium bikes for manufacturer Seven Cycles; by night, he runs a whimsical 80-member bicycle league called SCUL (for Subversive Choppers Urban Legion). This ragged crew is generally made up of high school teachers, engineering geeks, professional bike messengers, and other young locals looking for a weekend outlet. Every Saturday night, members begin their cruise through the city by picking from any of the dozens of modified bikes in the fleet, some as high as 18 feet tall. Of course, Skunk's is heavy, elaborate, and utterly terrifying to watch coming down the street: the USB *Catastrophe*.

Skunk, 36, began with the frame of a classic Western Flyer bicycle, circa 1938. He machined a new front end and chopped out longer forks from rolled steel tube to get that lowrider look. On the handlebars, he mounted a scavenged residential AC breaker box, which controls the bike's lights, junkyard Alpine stereo, and FM radio transmitter. The transceiver originally came as a bag full of transistors and instructions, bought off the internet and requiring a full day to assemble with a soldering iron and help from SCUL engineers. A couple of 12-volt, 7-amp-hour sealed lead-acid batteries from a hobby store power it all, and they usually last the duration of the club's two-hour cruises.

Most of the work is done in SCUL's top-secret hangout, which bears the appropriately Tom Sawyerish name "Fort Summer." There, members have access to drill presses, vises, and bike forming equipment. For more serious fixes, Skunk's employer gives him full access to a TIG welder and lathe after hours. If he can't find a use for a discarded sprocket or titanium tube, he'll weld it into his art, a well-received sculpture series called *80Grit*.

Just how long do Skunk's welds hold? *Catastrophe*'s cycle computer has logged 13,000 miles, though there have been a few repairs since it went into commission in mid-2000. (There's even a soldering iron in the bike's storage bin that works off onboard power if anything goes wrong.) The biggest surprise to Skunk is the success of his FM radio transceiver as a source of inspiration to other riders. Most club bikes now feature FM car decks and speakers welded to their frames. "It's like sitting in a giant stereo system where all the speakers are moving around each other," says Skunk. "And in the center of the speeding bikes you get amazing Doppler effects."

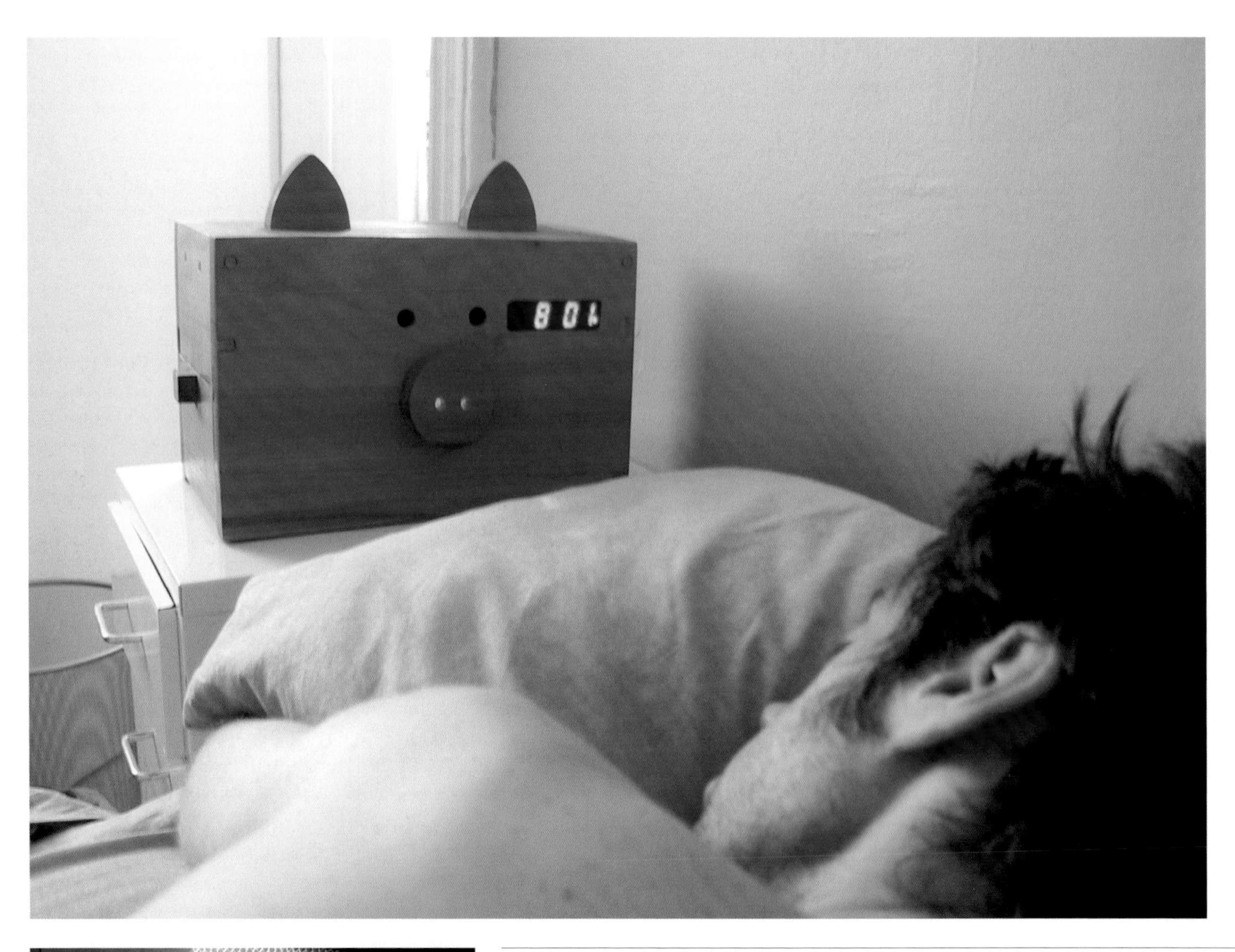

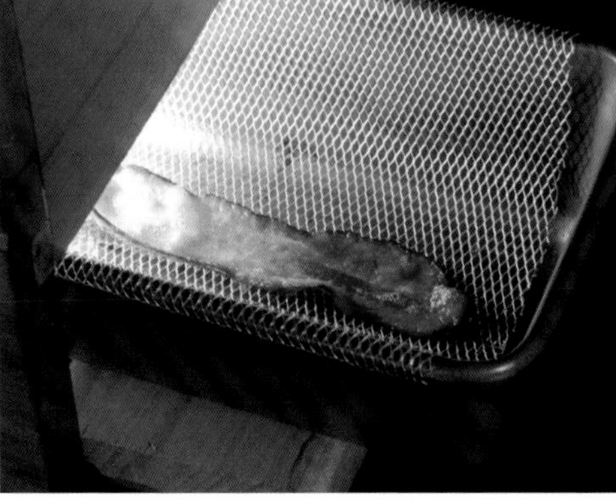

Above, the alarm clock cooks a piece of bacon to wake a slumbering user (the builder's friend). For tests, Sallin places a strip of frozen Oscar Mayer bacon the night before in a pull-out, foil-lined tray (left). When the alarm goes off, two 100-watt halogen bulbs start cooking.

Best Part of Waking Up

Bacon-cooking alarm clock

MAKER	**Matty Sallin**	COST	$90
OCCUPATION	Interaction designer, student	TIME	5 days
LOCATION	New York, New York	MORE	mathlete.com

Growing up, Matty Sallin spent summers in a rural Idaho cabin. He lived in Los Angeles during the rest of the year, but his parents hoped he and his sister would "appreciate the country" from the experience. For one thing, he remembers rolling out of bed for country breakfasts on Saturday mornings.

When he first powered on his Wake n' Bacon alarm clock — which cooks a piece of bacon instead of sounding an alarm — everything came back to him again. "The big sense memories are bacon, waffles, and cinnamon rolls," says Sallin, a web designer for top advertising agency R/GA and a grad student. "When I used it to wake up, my first thought was 'Mom's in the kitchen,' soon followed by 'The apartment's on fire.'"

Sallin, now 33, says he can't believe his "ridiculous contraption" actually works. It cooks the meat by activating two 100-watt halogen lamps inside the wooden case when the alarm goes off. He created it for fun over the summer as an improvement on a similar bacon clock he designed for a class project. The beginning electronics student takes circuit-building classes at New York University's Interactive Telecommunications Program.

Wake n' Bacon is constructed from a gutted Wal-Mart alarm clock and a PIC microcontroller. The chip receives the alarm signal from the clock, activating a series of relays to bring the electrical load up to regular wall current for the lamps. All the regular controls for the inexpensive digital clock are intact; because of cooking time, users are encouraged to set the alarm ten minutes before they plan to get up, allowing time to enjoy the smells and sizzling sounds.

Sallin is good at creating the consumer products you wish existed. He hopes to be a designer who bridges software and product design at a company like Apple or IDEO. One art installation he created, called ChitChat, uses a video projector to show graphical bubbles above the heads of two people in conversation. As the people talk, microphones in strategically placed beer cans pick up their voices, which are processed for volume and intonation. The person droning on the most earns a bubble that reads, "Blah, blah *blah*."

It's a little surprising that this student has decided to start at the bottom of a new field after attaining such high standing in web design. For five years, he worked as a creative director and lead programmer for such companies as Harman Kardon, Mattel, and Levi's. It was the little "sleep" light on the Macintosh computer, however, that inspired him to create projects that interact with people in their daily lives. "It delights people and has a little bit of soul," he says.

Many of Sallin's projects are very high-touch — a lamp that captures your heartbeat and then slowly throbs when you pick it up, an internet-enabled picture frame that lights up when his sister in California walks past. But the most attention so far has come from Wake n' Bacon, dismaying its designer just a little. Everyone loves it, even the vegetarians. "I always point out to them that you can put Fakin' in there," he says. "Though it probably won't smell as good when you wake up."

Fake Mac

Handmade Apple "G5 Cube" computer

MAKER	**Hideo Takano**	COST	¥3,700 ($33)
OCCUPATION	Toy designer	TIME	20 hours
LOCATION	Tokyo, Japan	MORE	conf.co.jp/pages/first.html

Like the unicorn or the jungle liger, the Apple G5 Cube is an animal that does not exist in nature. Computer aficionados have longed for such a machine, since the brief appearance of the beautiful but troubled G4 Cube, a machine Apple famously discontinued in 2001 because of a marketing snafu. The company has never mentioned bringing back the design. As one admiring fan wrote online: "It was an 8-by-8-by-8-inch masterpiece of space efficiency ... but it was destined to fail."

An admitted member of the worldwide cult of Apple, Hideo Takano, a toy designer in Tokyo, decided to create his own G5 Cube instead. He meticulously fabricated an outer shell mimicking the shape, styling, and paint job of one of Apple's latest, bulky G5 tower computers. Then he slipped this cover over his old G4 cube and happily continued working.

Takano fell in love with Apple's designs while looking for his first computer. "I'd think about it with the feeling of, 'Sometime I'd like to drive a Mustang,'" he writes in an email translated into English. "I swore to myself I'd definitely buy a Quadra 700. I think of those times nostalgically." More than a decade later, he still loves Apple, but he's hesitant to get the latest upgrade for design reasons. "Even if you were able to buy a G5, it's too big," he says. "The original excitement would wane once you owned it."

To start the project, Takano took measurements of a friend's Apple G5 tower and drew his own scaled-down design on paper. He then cut up ABS plastic sheets from a Tokyo modeling shop and glued them into a box shape. To get the metallic texture of the exterior, Takano covered it with a "sugary metallic powder" and painted it in silver acrylic. The logo required masking, a light coat of the metallic powder, and a contrasting paint job.

Takano created the power button with a piece of clear acrylic engraved with the "ON" symbol, then carefully placed it on the housing so that it would boot the computer and conduct light from the G4 inside. "Of course it slowly goes on and off, as though breathing," he says. "If I can't reproduce that kind of part faithfully, then I've failed at my craft."

To enable the front data ports, Takano simply embedded a store-bought USB hub into the new case. "I am not particularly focused on the electronic aspect," he says. "There are several Japanese sayings like, 'If you mix blood with something else, you just get red,' and 'You get good at the things you like,' and 'It is better to get used to something than to learn it.'" Good advice even for Apple execs.

The Solar Death Ray generates up to 1,000 degrees Fahrenheit thanks to 112 mirrors aimed with mathematical precision. Victims of its power include a burned-but-edible pan of Jiffy Pop (far left) and a lightly toasted Motorola cellphone (near left). Though his site, the builder actively solicits suggestions for future targets.

Live and Let Fry

Solar concentrator

MAKER	**Louis Giersch**	COST	$200
OCCUPATION	Aerospace engineer	TIME	40 hours
LOCATION	Seattle, Washington	MORE	solardeathray.com

Twinkies turn into fiery black husks in a few minutes. A Clue board game bursts into flames. A Richard Simmons video melts into a sticky plastic glob. As a concept, it's funnier than a Letterman stunt — and based on real science. Readers of the Solar Death Ray website can suggest any object at all, and Louis Giersch, 26, will obligingly expose it to 1,000-degree temperatures in a homegrown, 112-mirror solar concentrator.

The site's highlight may be its hilarious post-ray write-ups, which read like mock lab notes, but it was Giersch's attention to detail that made the Death Ray work at all. Giersch, who studies plasma spacecraft propulsion at the University of Washington, started by writing an application in the MATLAB visualization language. Mirrors on the 4-by-6-foot backboard have a tolerance of just 0.5 degrees either way to properly focus light, and the software supplied the exact angles Giersch needed.

He built small wooden bases for each mirror with the correct corresponding angle, and used a compound miter saw with an onboard protractor to cut the wooden blocks quickly. The blocks screwed into the frame along eight concentric circles from the plywood's center. (To make the circles, Giersch designed a large wooden protractor that anchored to the center of the board.) Mastic adhesive helped glue on 3.5-inch mirrors, hand-cut from cheap drugstore mirror glass.

Giersch — now Doctor Giersch — just finished his Ph.D. in aerospace engineering, a major hurdle in his ultimate career goal, to be an astronaut. "A Ph.D. is one of the minimum requirements, though there are thousands of applicants for every position," he says. "It's a real long shot." Before applying, he hopes to start a research career in his area of specialty, plasma propulsion for advanced spacecraft. "If I can't be an astronaut, to be a prominent rocket scientist would be great, too," he says.

In the meantime, he's working to bring the Death Ray to schools to curry interest in physics. Describing himself as a "big Lego kid," he says he grew up in Colorado buying disparate parts in the local hardware store, not knowing how to carry through his building plans. His parents were not involved in science or technology. "I had a definite need to make things, but no good ideas," he says.

An improved Death Ray is currently in planning, boasting a sun tracker and motors to automatically position it for the greatest solar blast. He'd also like to hook up a Stirling engine, a device that would use the heat to move a sealed piston. It sounds like the ultimate weapon of mass destruction, but not for use in Seattle maybe. "I've stored up a whole bunch of stuff that I really want to burn," says Giersch. "I have a Rubik's Cube all ready to go, but clouds keep coming up out of nowhere."

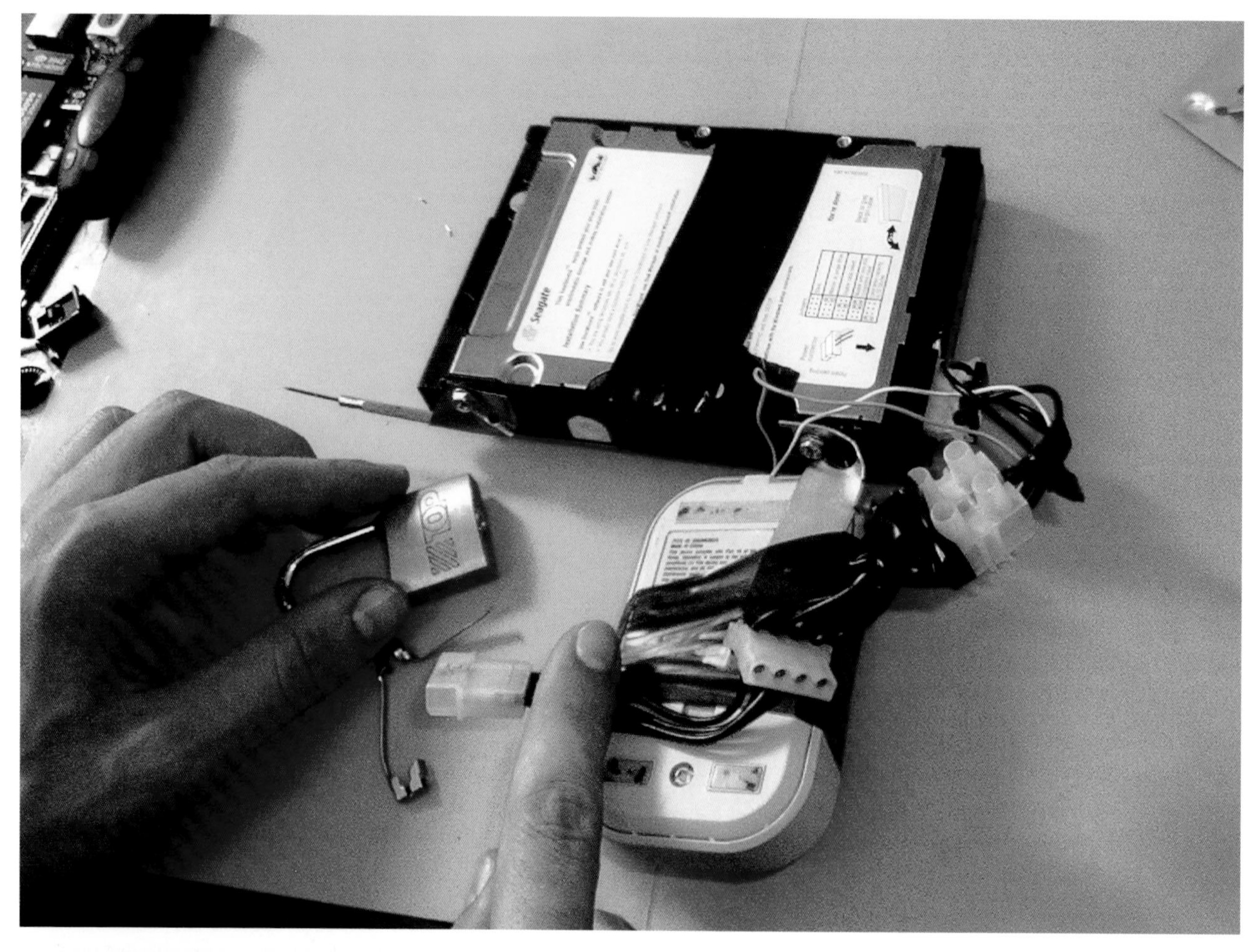

Another helpless padlock about to be cracked by the improvised electronic pick (top). From left to right, the hacker collective consists of DrDoom, Necro, and Klmar.

Pick of the Litter

Electronic lock-picking machine from castoff computer parts

MAKER	**David Martos, Bienvenido Melendez, and Jacobo Velez**	COST	$0
OCCUPATION	Computer technicians, textile industry operator	TIME	3 hours
LOCATION	Tordera, Spain	MORE	sorgonet.com

"The challenge was to build a lock-pick device by using computer parts. Of course we did it, and it works." Thus begins a how-to on the site of secret Spanish hacker collective SorgoNet. Well, maybe not that secret. If you instant-message David Martos Frigola, he'll tell you all about three friends who have worked on computer projects together since childhood. There's Martos, 30, who hacks under the pseudonym "DrDoom," Bienvenido Melendez Casado, 34, who goes by "Necro," and Jacobo Velez Masero, 30, who is "Klmar" online. They're perfectly nice guys, and they've built a closely watched site and a swaggering persona around some big hardware modification coups.

SorgoNet's mods are always low-cost with a touch of comedy. They filled a computer case with sunflower oil as the basis of a waterproof outdoor wireless network. An old hard disk became a stereo speaker in another experiment, and a crude CD-ROM player once made its way into Velez's subcompact car. Many of their projects are presented in both Spanish and English. ("I learned English by reading websites and playing games," notes Martos via IM. "My vocabulary is large in many areas, like medieval weapons and spaceships.") The trio typically messes around with gear in a 12-by-12-foot room in Velez's apartment until midnight on weekdays.

The lock-picking hack took place on a lazy Sunday afternoon, and it reads more like a piece of performance art than a useful device. It's similar to a locksmith's common vibration gun, in which a pulsing metal pick nudges pins inside a lock, allowing the cylinder to turn. SorgoNet's project was able to open a small household lock in a few seconds, but it was tethered to an AC wall adapter, and held together with electrician's tape — hardly something you'd want to depend on for that big embassy break-in.

The three gadgeteers first scrounged up a broken 7,200RPM hard disk and removed the read head and steel plates. Then, they turned the drive into a vibrator by bolting an off-center piece of steel to the motor shaft, causing the whole thing to shake. A small screwdriver attached to the outside of the hard disk picked up the vibrations and acted as the lock pick. The group then dissected a three-button mouse, using its switch to fire the device and its body for a pistol-like grip.

According to Martos, hacking provides a fun way to rebel against products straitjacketed into single, boring purposes. "Everyday technology has become more and more sophisticated," he says. "So when somebody can hack something, it's really nice. Things aren't as complicated as they seem." He and Velez work in small computer shops reinstalling Windows and removing viruses. Melendez operates machines at a textile manufacturer and has developed a role-playing game in his spare time (*black-legacy.com*). Velez and Melendez left school at 16 to pursue their trades.

Even when away from SorgoNet HQ, it's never difficult to round up a fellow hacker. The friends' apartments triangulate 250 yards around the wireless access in their windows and roofs in this small town north of Barcelona. The home-built Linux systems allow them to trade files and play Quake from the comfort of their respective living rooms. All three men are single. Says Martos: "Only singles have time to do this geeky stuff!"

Living on a Thin Line

Computer-controlled laser cutter

MAKER	**Owen White**	COST	$15,000
OCCUPATION	Laboratory director	TIME	600 hours
LOCATION	Rockville, Maryland	MORE	nilno.com/laser_intro

Heard about advanced tech that will let you print out common household objects faster than glossies from an inkjet? It may be causing a buzz right now in the press, but to Owen White it's yesterday's news. Since February 2005, White has been pumping out custom objects in his basement from ⅛-inch stainless steel, using a computer-controlled 100-watt laser. Whatever the 44-year-old bioinformatics scientist can sketch out on his computer screen, he can have in hot metal a few minutes later. He's made sculptures, jewelry, even support brackets for a drum stand.

"It's like when Galileo made the telescope — he basically could point it in any direction and make a discovery. Pretty soon everyone will have one of these, and it'll be as common as an iPod," says White.

White had to wait quite a while for the privilege, checking eBay five days a week for three years, looking for a suitable 100-watt CO_2 laser within his budget (the devices can retail for as much as $28,000). One day, he saw one listed at $6,500 and couldn't open his PayPal account fast enough.

White, who runs a lab of 50 scientists at The Institute for Genomic Research, has always had an interest in sculpture and modeling. Having spent hours cutting out intricate pieces by hand for radio-controlled airplane projects, he wanted a table that would automatically cut materials based on computer commands. For a year and a half, the laser sat on a shelf while White designed the electronics that would control it. With a casual background in electronics left over from high-school projects, he used Proteus and Eagle circuit-design software, which make it easy to grab sample designs off the net and modify them for your own needs.

Then White sent the design, built around an Atmel microcontroller, to a service bureau in Korea, which charged only $65 for a final circuit board. He ordered the steel parts for the table from a machine shop, and MIG-welded some I-beams together to make a stand for the oxygen tank (when the laser vaporizes metal, little blasts of oxygen clean the cutting surface). The laser is fixed above a platform that moves underneath, controlled via serial cable to a rather ordinary 1.2GHz Pentium IV machine running Unix. White designs objects in CAD, then uses the OneCNC profiler to convert the 3D image into a CNC-friendly script called G-code. Another app turns G-code into EMC, a signal that goes out the printer port.

Now that he's worked out the kinks, White is eager to start collaborating with his wife, Imani, who designs glass beads out of a workshop in their home. The table produces flat pieces, but White rivets them together or bends them with pliers for skeletal likenesses of frogs, birds, and fish. Imani wants to create the eyes and glass features for the animals.

White says it's still "a hoot" to create objects in his basement, though not as exciting as that first eureka moment several months ago. He had been frustrated trying to get the device to cut when he made some adjustments to the oxygen pressure. He saw a spark under the optics, and picking up the stainless steel, saw a tiny hole in the metal. "That's when God spoke to me."

i'm almost famous!

Art Machine

Computer-guided magic marker

MAKER	Tristan Perich	COST	$150
OCCUPATION	Student	TIME	2 days
LOCATION	New York, New York	MORE	tristanperich.com

Call it minimalism via HAL. Tristan Perich's huge wall murals are really made by a computer; he just nudges the artistic process along a little bit. Perich, 23, set up his art-making machine using a couple of stepper motors, a laptop, and a black Sharpie. With that simple apparatus, he's drawn abstract images up to 14 feet square in places like art studios and urban staircases.

Perich starts out with a black-and-white photograph, which he feeds into his PC. The computer sends data about the image to a circuit board with two stepper motors. He fastens the motors to the top left and right of the wall space, then their small movements pull at fishing line, causing the Sharpie to draw.

Although there have been similar DIY wall printers in Germany and the U.K., Perich is more interested in making non-representational art. The laptop he uses is programmed in Visual Basic to send data along a serial cable about the light and dark areas of a photo. Then an Atmel microcontroller composes the wall image according to a few simple rules: light areas produce straight lines and dark areas produce squiggly ones.

Perich says the wall pieces remind him of the structured, minimal works of one of his favorite artists, Sol LeWitt. (He grew up looking at some of the top minimal artists at his mother's business, Candace Perich Gallery.) In subject matter and technique, he's also following in the footsteps of his dad, Anton Perich, a former contributing photographer for Andy Warhol's *Interview* magazine and creator of an innovative electronic printing process in 1978. Perich senior developed a large print head on a steel scaffolding. Loaded with hundreds of colors, it draws horizontal lines in various colors to generate abstract works. But where his dad's images were literally dripping with colors, the son's are monochromatic, mathematical, and a bit detached.

If you saw Perich on the street, you might think he's a typical Brooklyn scenester, clad in jeans and fitted tees and clutching a beige household phone that's been modded to make cellphone calls. You might be surprised that he's an accomplished pianist and a recent Columbia grad with degrees in both math and music. Or that the current focus of his life is hand-soldering circuit boards. He recently started classes at New York University to study electronics and art, and he approaches making electronics like it's his great calling.

The phone, for instance, was one of his first soldering projects, in which he connected the keypad for the full-size phone to an LG cellphone buried inside. Because of mistakes with hot solder, he first ruined eight cellphones (they came very cheap if you knew how to work the AT&T upgrade system, he says).

After learning a simple oscillator circuit, Perich created a crude music player that will sell in select New York record stores in January 2006. The "One Bit Music" device looks exactly like the jewel case for a CD, except that it has a little jack on one side. Connect it to headphones or any amplification system, and it plays simple electronic dance tunes.

He'll certainly get plenty of soldering practice. Perich wants to handmake 800 of the units over the next few months, although he says slinging hot tin still challenges him, like a particularly difficult piece on the piano. "Soldering would be so much easier if I had three hands," he says. "As it is, I have trouble trying to hold the two wires and apply the solder at the same time. My technique is awful."

Horse with No Name

Motorized, walking table

MAKER	**Douglas Repetto**	COST	$200 plus salvaged parts
OCCUPATION	College professor	TIME	2 months
LOCATION	New York, New York	MORE	music.columbia.edu/~douglas

"The foal does not fall down!" says Douglas Repetto in a tone that makes it both a declaration and a warning. Repetto built the creaky walking table to mimic the first movements of a baby horse, and the art piece is due to make an appearance at a prestigious exhibit in Switzerland. Repetto doesn't want any hitches. "Over the last few weeks, I've been working hard to understand why it would fall down," he says, then adds a bit defensively: "It hasn't fallen in a long time, by the way."

If performance anxiety is getting to him, it's not surprising: the mild-mannered 35-year-old is usually behind the curtain, not sweating the details out in front. Many people know Repetto as the impresario of the popular Dorkbot presentations in New York City, which bear the tagline "People Doing Strange Things with Electricity." He started the geek show-and-tell sessions in 2000, and the idea has since spread to 28 cities around the world. Repetto, who teaches computer music at Columbia University, has made dozens of electronic art projects, but he's increasingly fielding requests to show them in the U.S. and abroad.

The ordeal began when the Festival Rümlingen in Switzerland asked him to present an artwork in the middle of a grassy field. "I started wondering what would happen if radiation hit a table one day and it suddenly walked away," he says.

Repetto first studied foals in birthing videos from farms to see how they made their stiff-legged entry into the world. He gathered scrap two-by-fours from the wood shop at school, and ripped the boards lengthwise to make the legs and structure. (His office is housed in the same building as the visual arts department, which has a fully stocked wood and metal shop.) He then nailed a piece of laminate on top. With a milling machine, he cut a slot into a round steel disk to make a drive system for the legs. A wooden peg in the metal slot creates the lateral movement while the motor turns. He mounted four scrap office-machine motors and powered them with a 12-volt motorcycle battery. An Atmel microcontroller picks a new, arbitrary speed for each leg every few seconds.

The foal isn't supposed to be a robot; it has no sensor inputs and no decision tree. "You could write a lot of code to precisely control the motors, but I wanted to find out the simplest way to create interesting movement," Repetto says.

The foal's brutal, lurching stride may be interesting, but it's not elegant. All four of its legs typically stay on the ground while it pushes to and fro on polyurethane casters. "It is just a table wildly flailing its legs but you immediately get sucked into it. You start reading all this intention and desire and struggle," says Repetto, who often consults his wife, writer Amy Benson, as a sounding board for what works. "She doesn't have a background in tech, so she doesn't get caught up. You can get trapped into thinking that doing something difficult is the same as doing something interesting — and it's not."

Repetto's classes at Columbia focus on making electronic music systems and introducing electronic sounds in artworks. He joined the faculty in 2000, after teaching similar topics at Dartmouth. With two Ivy League posts on his resume, you'd think he might have a stellar academic background, but school didn't engage him until he went to art school at the California Institute of the Arts. He says his undergraduate work in music composition and film at the University of Miami »

Above, computer-controlled lights cause a houseplant to slowly zigzag from left to right. The result is shown continuously on a sped-up video feed. A time-lapse video (bottom, left) reveals the daily opening and closing of a prayer plant. The artist's workbench (bottom, right) is located at Columbia's Computer Music Center.

The 2005 piece ***Puff Bang Reverb*** **recreates the effect of molecules moving through air. When a viewer blows lightly into an air-sensitive switch, two motors spin in the center, creating ripples through 600 wooden blocks suspended by fishing line. Here, the artist's wife helps put the finishing touches on an exhibit near Basel, Switzerland.**

» was not distinguished. (He once noted in an interview: "I was a horrible, horrible student; I thought I was going to be a manager at Blockbuster.")

Finding like-minded artists around Dartmouth's New Hampshire campus was tough. As soon as Repetto landed his current job, he started formulating a way to bring together tinkerers and electronic artists from among the non-student population in New York City. He named the first meeting "Dorkbot" to keep its intentions humble, and encouraged people to bring projects made on the cheap out of "obtainium" — whatever surplus or thrashed gear they could find. In 2002, he started ArtBots, an annual talent show for robots, as an antidote to testosterone-fueled events like Robot Wars. And he runs a well-regarded blog called *Organism: Making Art with Living Systems,* where he frequently comments on how technology has given artists new ways to make biologically based art. In his own work, this translates into computer-based projects that allow him to annoy houseplants and display their movements on video screens.

Although the foal does not have the least bit of electronic intelligence, people who've seen the table tend to muster a lot of sympathy for it. "You really get the sense that this thing cannot do what it's trying to do. Your heart just immediately jumps that you want to help it," he says, and then pauses and laughs. "It's especially emotional for me because I'm thinking, 'Oh, crap, I'm going to be without my tools in the middle of a cow field in Switzerland.'"

Tilman Reiff (top) sets up the punching mechanisms for the wooden puppets. An Atmel microcontroller handles scoring, lights, and movement of the velvet curtain at the start of each game. Mr. Punch debuted at the Rohkunstbau museum near Berlin (near left).

Mr. Punch

Mechanical boxing game

MAKER	**Tilman Reiff and Volker Morawe**	COST	€1,600 (about $1,900)
OCCUPATION	Multimedia designer, electrician	TIME	3 months
LOCATION	Cologne, Germany	MORE	fursr.com

The homemade arcade games of Tilman Reiff, 34, and Volker Morawe, 35, always look good in the workshop, but the designers can never judge their final success until people push buttons and scream at them. When visitors started playing with Mr. Punch in its quiet gallery setting, the two knew they had struck it right.

"The reactions were too good," says Reiff. "They were smashing against each other — *ha-POMB, ha-POMB* — and the machine only going *ding, ding, ding*. I was panicked, saying, 'No, please be careful.'"

Mr. Punch is a simple, elegant contraption in which players pit antique doll heads against one another, causing them to jab with a bicycle lever, and keel over in half. Its creators, who met in a master's program at the Academy of Media Arts, Cologne, are the undisputed kings of retro interactivity. Perhaps their most famous work is PainStation, a video game console in which the simple game of Pong is made painfully real when a mechanical lash emerges from a panel and literally whips the player after a lost point.

In this context, Mr. Punch is fairly reserved, a throwback to a simpler era of gaming. It's immediately made museum-goers fall in love with it, perhaps a little too much. Mr. Punch was developed for an exhibition located near Berlin, several hours from the designers' houses. After a few overzealous players loosened some of the parts, an elderly guard at the museum voluntarily took it under his wing. "He began putting grease on it, and checking to see if everything's working," says Reiff.

To start the project, Reiff and Morawe dug around for metal parts. The puppets travel along steel rails taken from a filing cabinet, and are moved laterally by bicycle frame tubes. The puppets themselves are from eBay, but to create the boxing gloves, the men went around to local bars collecting wine corks, then whittled them down to fit. They designed the punching mechanisms in a vector graphics program, sending them out to be machined in thick sheet metal.

To throw a punch, the player grabs a bike handlebar and operates a lever taken from a bicycle brake. This connects to the doll's mechanism via a common bike cable. Reiff, who speaks immaculate English, grows momentarily glum. He can't think of a suitable translation for "brake cable" and can only offer the words "brake cable." Apparently, there's a much longer term in German with unparalleled precision.

To make a living, the two perform decidedly more mundane forms of interactivity— Reiff builds websites in Flash, and Morawe is an electrician for a theater. They earn enough from their museum work to pay rent on a small storefront they keep, calling their two-man workshop "Fur." Morawe left school early to apprentice to an electrician, and Reiff hung on, eventually getting a degree in computer programming. But he says he was always fairly bored: "I had a lot of disciplinary problems, always drawing caricatures of my teachers during class."

Reiff says his favorite aspect of Mr. Punch is the way the electronics are hidden from the user. When a new player presses "Start," a screen opens to reveal the interior, and when a puppet is knocked own, a small pinball flipper rights it. The designers drew from a cache of vintage pinball components to create the analog score wheels, which turn in response to servos and an Atmel microcontroller. Says Reiff: "I love the sounds they make when the score changes. When the game starts, it resets to zero, *CLACK, CLACK, CLACK, CLACK*."

The Cuisinart Whisperer

Voice-activated blenders

MAKER	**Kelly Dobson**	COST	$300
OCCUPATION	Researcher and student	TIME	300 hours
LOCATION	Cambridge, Massachusetts	MORE	web.media.mit.edu/~monster/

Most people swear at their kitchen appliances from time to time. But Kelly Dobson, 34, actually speaks their language. She bought a dozen old blenders on eBay and is rigging them up to change speeds based on voice commands — very peculiar voice commands.

To get them to work, Dobson, a doctoral candidate and researcher at MIT, imitates the blenders with low, guttural motor sounds. "You have to make sure you have the roughness of a blender, so you might get him started with a *rrrrrRRRRRRR*," says Dobson. When she raises the tone, they speed up. Because each one has a slightly different mechanical voice, Dobson has trained them to respond to particular tones.

Dobson is very soft-spoken in conversation; you have to lean in close to catch every word. But when she talks about her family of blenders, she's more animated. (Friends have given her the nickname Monster because of the strange *grrr* sounds she uses in her research.) She entered MIT as an artist, but her work has dimensions of both art and science.

Getting the blenders to know when you're talking to them is no mean feat. Dobson is an expert in digital signal processing, and a lead author of an October 2005 paper called "Learning Auditory Models of Machine Voices" presented to the IEEE. She programmed algorithms in C++ into 1GHz VIA Epia microprocessors mounted inside each blender. The processor uses short-term Fourier transforms, a way of breaking down the sound into chunks of time. That way the blender can listen for vocalization changes, and better judge whether you want it to speed up or slow down.

For the October research paper, Dobson worked with two other engineers to train a roomful of hardware — sewing machines, drills, blenders — to respond to different voices. Though they analyzed instantaneous sounds, not the more complex Fourier transforms, the machines responded appropriately 60% of the time. The research could one day allow for better voice control of multiple devices in a home or factory.

Dobson's work is a step forward in auditory modeling, but her main goal is a cultural one — to debunk the myth that humans will ever have seamless relationships with machines. *Star Trek*'s Data not withstanding, the artist wants to show that machines can be a little bit neurotic or even rebellious. In her Boston apartment — which she shares with her computer scientist boyfriend and a workbench full of oscilloscopes, power supplies, and soldering equipment — she says her blenders don't always behave. "Sometimes they'll go on when you start talking on the telephone," she says. "Or they'll misinterpret laughing as someone talking to them."

Growing up in Detroit, Dobson had a wild early introduction to hardware. Her grandfather staffed a junkyard, and her parents let the four sisters and brothers roam the lot freely, smashing windows and poking around under hoods. "We were supposed to be under the supervision of my dad, but he wasn't very good at it," she recalls. "They gave us so much room to do our own thing. I think that's good for kids."

4-40 NUTS
CAP-HEAD 6-32 x 5/8
CONT. ROTATION SERVOS
HI-TEC SERVO HORNS
CAP-HEAD 6-32 x 3/8
DPST SWITCHES
IRLZ34
20MHz RESONATORS
4MHz RESONATORS
THREADED 6-32
NYLON STANDOFFS 3/4
6-32 NUTS
STD-HEAD 4-40 x 1/2
9V BATTERY CLIPS
40-PIN SOCKETS

Parts Mart

Automated parts dispensing machine

MAKER	**Zach Radding**	COST	$1,000
OCCUPATION	School director	TIME	200 hours
LOCATION	Berkeley, California	MORE	buildcoolstuff.com

"You learn better when you can bang on something and wiggle the wires," says Zach Radding, 30, the director of a small electronics and robotics school in Berkeley, Calif. It's a credo that he preaches to his students and practices himself. "I always dive in and build a prototype, come what may."

In this case, the teacher's latest masterpiece is an automated, PC-powered organizer for the school, a 12-by-6-foot shelving unit with almost 400 different bins for things that students will need.

Radding's school, called Zach's Cool Stuff, is for kids and adults, offering everything from the construction of simple robots to advanced electronics prototyping workshops. Usually, students have to dig around for such oddities as Hall-effect sensors or cadmium sulfide cells. But next semester, students can choose parts from a menu on one of the desktop PCs around the shop, and the correct parts will be presented to them by the time they walk by the bins.

Radding started by constructing a simple shelving unit out of medium-density fiberboard and steel supports. He created a motorized platform on rails which moves left and right, up and down, looking for the correct bin. When it hits the jackpot, it grabs the bin via a hook on a lead screw, and pulls the bin onto a tray.

At first, teachers at the school brainstormed how to make sure the platform knew its location. They briefly considered servomotors which would index their place among the bins, but the shelves might warp and subtly change position. Instead, they installed small LEDs and optical pickups on the platform, using the plastic bins themselves to reflect light to the sensor; the platform only needs to count the number of light blips vertically and horizontally to get to the right bin. "We designed it so it can be kinda sloppy, but sloppy's OK," says Radding.

The organizer is still a prototype, but one that works remarkably well. The same could be said about Zach's Cool Stuff, which attracts 30 to 50 local students per month in after school and weekend classes. Radding paid for the place by rehabbing Oakland townhouses in a hot real-estate market in the late 1990s. Before that, he had sort of bounced around the halls of high tech, not exactly fitting in. His first job was fixing Unix hardware at banks and corporations for HP; then he went to art school, dropped out, and became a robotics builder at Lawrence Berkeley National Laboratory.

This latest role seems the best fit. "I'm doing it all on my own. It's a lot more flexible and fun to make things in your own place."

Private Jet

Custom flight simulator

MAKER	**Hans-Joerg Krohn**	COST	€10,000 (about $12,000)
OCCUPATION	Bank manager	TIME	10 years, part time
LOCATION	Belgrade, Serbia	MORE	hanskrohn.com

He may be crammed inside a tiny apartment in Belgrade, Serbia, but 46-year-old Hans-Joerg Krohn takes flights to Hong Kong, India, and Austria on a nightly basis. When the plywood canopy slowly arcs to a close, Krohn monitors a seamless panorama consisting of three 21-inch high-res CRT screens across an ersatz windshield. Next to him, a motorized throttle follows the flight path on autopilot, while five separate navigation screens provide data, and hundreds of buttons flash on home-made instrument panels.

"I try to tell myself it's just a piece of wood sitting in a room," says Krohn in impeccable English with a slight German accent. "Something happens in the brain because of the closed canopy and the high workload inside. You have to flip switches, key in information. You experience complete suspension of disbelief."

Way back in 1995, Krohn started building the elaborate sim because he missed flying the real thing. A manager of an international banking department, he moves to a new country every three to five years at the behest of his German employer. After a relocation from Mexico to Kazakhstan, he knew his private pilot days were over. Individuals simply don't fly the friendly skies of Kazakhstan as a hobby, says Krohn.

In a U.S. magazine, Krohn saw a full-size F-16 cockpit replica. He bought it for $300, then filled it with his own electronics and dozens of custom control panels. Krohn's painstaking routine includes designing the panels in CorelDraw, then having them laser-cut in sheet metal at a local engraving shop. He prints the labels out on special foil paper, affixes them to the sheet metal patterns, and adds knobs in front and hand-soldered circuit boards behind.

All the electronics feed to EPIC boards, which are low-cost control circuits designed for flight simulation fanatics. They allow builders to create their own buttons and program them using the simple EPIC programming language. (Krohn makes all his scripts available for free online.) The boards and displays, in turn, are controlled by a special flight sim application called Project Magenta. The suite runs Microsoft Flight Simulator 2004 and offloads various functions to eight daisychained PCs ranging from a Pentium II 300MHz to an Athlon 3.5GHz.

The hardest part of the project was creating the 22-pound canopy from thin strips of plywood. To get the curve of the rounded top correct, Krohn improvised a system he'd read about in Spain, used on Antonio Gaudi's Sagrada Familia cathedral. He soaked pieces of rope in glue and let them hang in a semicircle until dry. These ribs created a structure for bending the plywood. Krohn attached pneumatic struts to each hinge, so the top wouldn't bang him on the head when it closed.

Just as in a real jet cockpit, some of the physical controls move on their own to keep the pilot oriented. Krohn modeled the joystick throttle after an F-16 fighter's, which moves when the autopilot applies thrust. The ghostlike motions help the pilot determine where the controls should be during an emergency. Krohn created his own 1-to-810 gearbox for the auto-throttle using nylon gears he bought on the net. He hooked up a servomotor and built it into a sturdy PC game throttle. For the brake, he created a feedback system based on a similar motor configuration, modeled after a system used in the cockpit of a Boeing 737. »

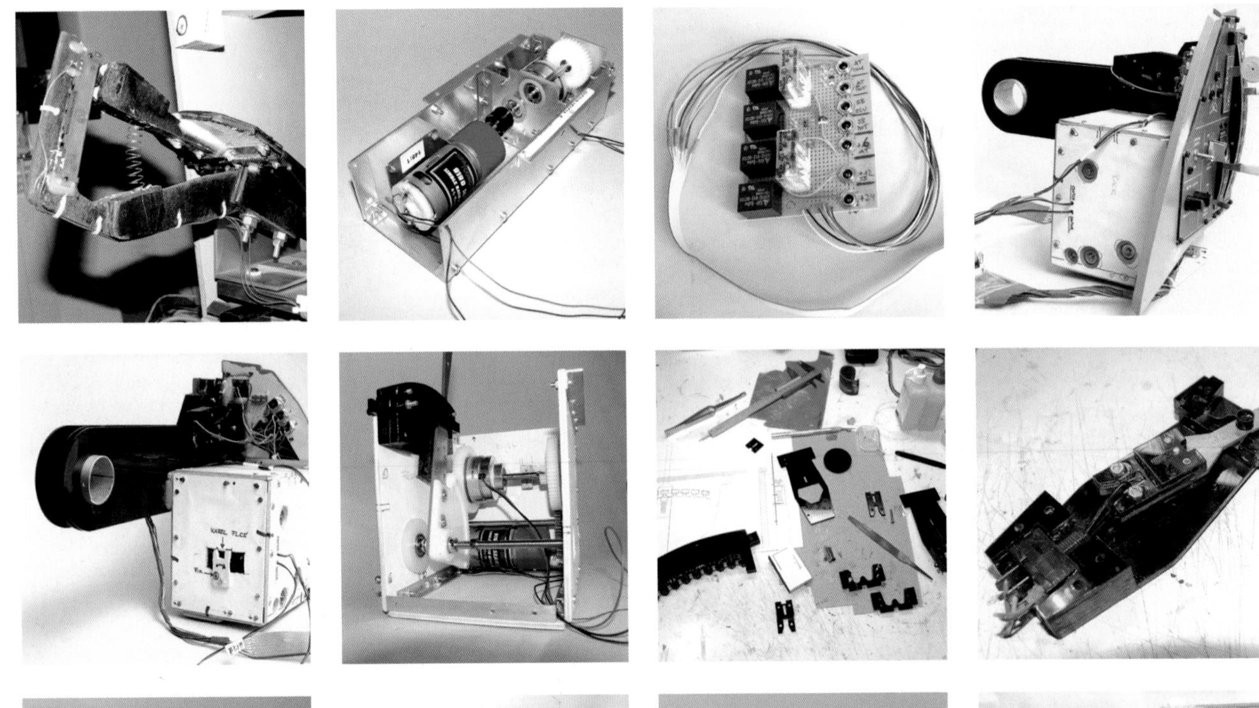

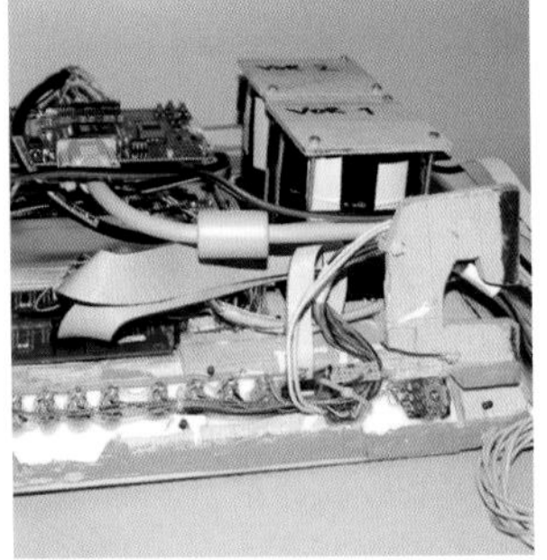

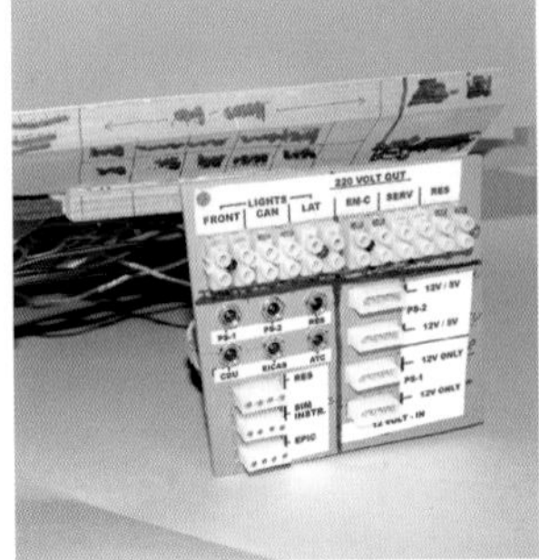
220 VOLT OUT
LIGHTS
FRONT
CAN
LAT
SERV
RES
PS-1
PS-2
CDU
EICAS
ATC
SIM INSTR.
EPIC
12V ONLY

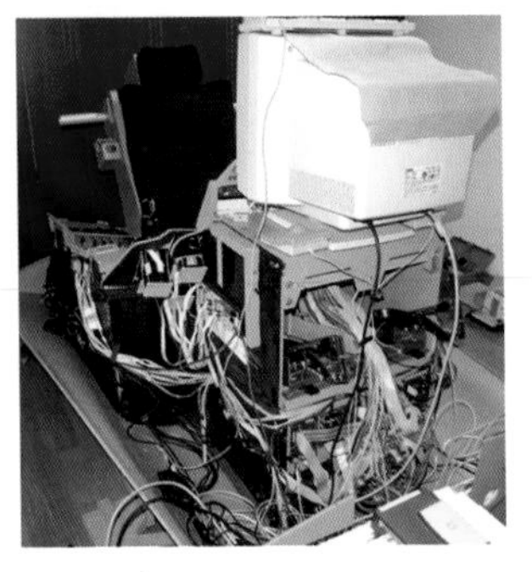

NAV
N
E
S
W
33
30
24
21
15
12

Above: All the instrument panels for the flight simulator are handmade from metal, paper, and acrylic. Five built-in screens show positioning data, pitch, roll, thrust, and time to destination. Eight PCs coordinate the flight data to the various displays and controls. Krohn's latest setup uses three 21-inch CRT screens, not pictured, atop the panels to show the virtual landscape outside the cockpit windows. The hardware is a mashup of jet fighter and commercial airline controls. Opposite: Construction details of the servo brake, auto-throttle, and analog standby instruments.

» The mongrel setup doesn't earn much status with other flight simmers, most of whom aim at meticulously recreating one craft. But that doesn't matter much. "This building process is something that gives me a lot of joy," says Krohn. "When something works, it makes me a happy person. Having the whole thing finished would be kind of an anticlimax."

When the bank transferred Krohn from Kazakhstan to Belgrade, he dragged the whole collapsible cockpit with him. His wife has tried it a few times, though she has only a minor interest, leaving Krohn to his hour-long flights a few times a week. "I was over India one Sunday afternoon and suddenly the phone rang," he recalls. "My wife knocked on the canopy and the light came in. Suddenly I had to talk about something very down-to-earth."

Packed with more than 1,200 optical fibers, the table's surface displays light and shadow from a grid of fibers placed on a nearby window. Dreary office denizens see the shadows of birds and trees moving outside, no batteries required.

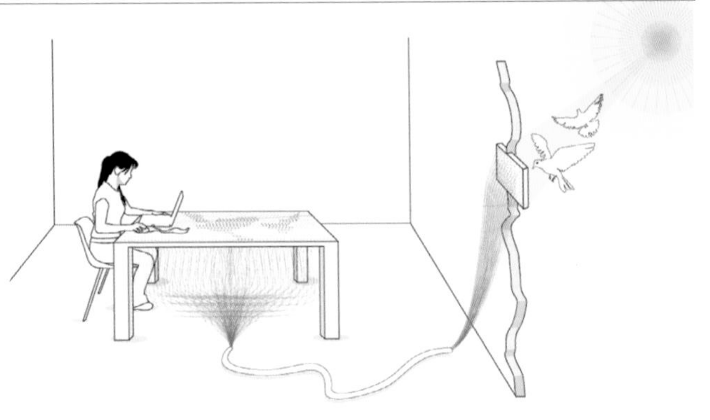

Holy Workspace!

Fiber-optic desk

MAKER	**Anab Jain and Stuart Wood**	COST	$3,500
OCCUPATION	Interaction designers	TIME	400 hours
LOCATION	London, England	MORE	anab.in, random-international.com

Cubicle land is as far from the great outdoors as breakroom coffee is from real java. The impetus for design students Anab Jain and Stuart Wood's light-up table came when they witnessed the awful conditions of dim, cramped call-center offices during a six-month research project.

Their "Sunlight Table" is the response. It uses 1,200 optical fibers to pump sunshine from an outdoor panel to the table's wooden surface. If a bird or tree branch casts a shadow over the panel, the outline is enlarged and reproduced on the desk inside. "We wanted to bring the outside in and give the feeling of natural light," says Jain, 28, who teamed up with Wood, 25, while they were completing their master's degrees in interaction design at the Royal College of Art in London.

Before building the piece, Jain researched call centers in Newcastle, England, and some closer to where she grew up. "A cousin of mine and her friend were working in a call center in India," says Jain. "They made her practice a British or an American accent. She gave up, but her friend has been working there for three years now. She doesn't see her family for the whole week. It's very lonely and depressing."

As it turned out, the table would have to be more of an inspired gesture than a real product for the notoriously cost-sensitive call-center industry; it was hugely labor-intensive. Jain sawed much of the wood herself, creating the parts from beechwood laminate in the school's workshop. Because of the overwhelming number of holes that needed to be drilled, the boards then went to a CNC shop, where computer-controlled equipment automated the drilling.

And then came gluing the optical fibers. "It was a painful process, a nightmare," says Jain, who earned an undergraduate degree in film in India. The designers stuck every single optical fiber into the 1,200 holes by hand, fixing them with super glue and silicone. The other ends of the fibers had to be correlated into a corresponding hole on the smaller panel, so that light patterns and shadows would reproduce accurately. One person would hold a flashlight to the fiber ends, while the other stuck them in. Superglue was unforgiving, and they made a few mistakes. "I started to have dreams of me getting entangled in the fibers. We were like a divorced couple toward the end," laughs Jain.

"There were lots of scrappy fights on whose turn it was to stick the fibers into the table," says Wood, who laughs good-naturedly, suggesting that we don't know the half of it. He was a product designer before entering the Royal College, and since graduation has started a design studio called Random International. He and a partner have developed an electronically enhanced device that looks like a paint roller. Called PixelRoller, it allows the user to paint giant photos and graphics on a flat surface as they are sent down a computer cable. The group is currently in talks to commercialize the device and can't divulge the technical details.

Meanwhile, Wood and Jain's table has been bought by a collector for almost $9,000, which Wood thinks is fair: "It has a lot of my blood on it." And of course, it's pretty cool. "You don't usually get the big 'wow' when you're the maker. Your nose is too close to it," says Wood, who sounds a little like chef Jamie Oliver. "But we did have one stand-back point where we had all the optical fibers in and turned it over, cut the ends off, and shined a light on them. We said, 'My God, it looks wicked.'"

Waterbug

Dual-prop pedal-powered canoe

MAKER	**Erhardt Wowerat**	COST	$700
OCCUPATION	Electrical engineer	TIME	80 hours
LOCATION	Milwaukee, Wisconsin	MORE	bikeforest.com/homebuild1.php

Somewhere in a basement in Milwaukee, a man has created a souped-up canoe that promises to run circles around wimpy paddleboats and swan boats. His invention uses dual propellers for speed and steering, a power train that sits high in shallow water, and most importantly, a system of cycle gears and derailleurs for varied speeds. "Bikes were once one-speed," says Erhardt Wowerat, 36. "Why can't we improve pedal-powered boats, too?"

His first try was not quite seaworthy. Wowerat made an unorthodox propeller out of wood that was supposed to jet water out the back through a heavy wooden channel. Unfortunately, this electrical engineer for a construction company was soon frustrated by the limit of his mechanical skills. "It didn't work at all," he says. "I didn't sink, but I wasn't going anywhere either."

For the Mark II version, Wowerat switched to aluminum, now his medium of choice. (He played with Legos as a kid, and working with aluminum reminds him of those days.) He cut the ⅛-inch gauge pieces using a table saw fitted with a carbide blade. Through trial and error, he learned how to weld it, putting a small fitting on the bottom of his aluminum canoe to bolt the apparatus on. He cannibalized parts from a Trek 1000 bike for sprockets, derailleurs, and a front crank set. He bought a machinist's miter gear to transfer the crankshaft's force 90 degrees to the propellers whirring out back. And the extravagant seat? "It's a tractor chair," notes Wowerat. "It has detachable armrests and adjusts forward and back. I saw it in the store and just said, 'That's the one I want.'"

Aside from a few run-ins with lake weeds, the rig works well (alas, his wife has yet to take it for a spin). Wowerat can't wait to tackle the next version. The son of German immigrants, he spent nearly every day of his youth boating around Lake Winnebago in Michigan. He's been taking night classes at Milwaukee School of Engineering to get his master's and improve his mechanical engineering skills. He recently bought AutoCAD software for his computer and hopes to test canoe designs before getting bogged down in a local lake again. His next version will be more modular, a Lego-like add-on to any canoe. Just think: one day you too could turn your gentle little boat into a killer watercraft.

Chitty Chitty — Bang! Bang!

Go-kart with turbine jet engine

MAKER	**Richard Flanagan**	COST	$1,000
OCCUPATION	Machinist	TIME	1 week
LOCATION	Lebanon, Oregon		

In the small town of Lebanon, Ore., you might see Richard Flanagan cruising an empty farm road in a jet-powered go-kart. But you'll definitely hear him. The 57-year-old machinist for local sawmills can reach speeds of 60 miles per hour and, more importantly, generate a lot of fire and noise. "People come out of their houses to look," says Flanagan. "At night, it lights up the neighborhood."

To hear Flanagan talk about it, jets are easy. Like other jet enthusiasts, he usually starts with a turbocharger from an old car, then adds his own shafts, ceramic ball bearings, and custom-welded chambers. (In an online buying spree, he's holed up 75 used turbochargers in a storage shed for future projects.) To keep everything running smoothly, he created an electronic throttle control, which captures sensor readings on exhaust temperature, rpm, and volume of fuel. The kart's driver only needs to turn a solid-state dial to make it go faster or slower. A car battery and power converter deliver the necessary voltage for the electronics and ignition.

Flanagan started his career making jet parts for the aircraft industry in Southern California, following the path of his father who worked for the Air Force's Air Research and Development Command. Now, he works as an independent contractor in the machine shop in back of his house. (His compressor flips on and off intermittently while he talks on the phone.)

After selling three karts on eBay Motors, Flanagan hopes to create more. After all, who wouldn't want a ride with push-button ignition and the option of using gas, diesel, or paint thinner as fuel? All for around $1,500. "The demand is incredible," he says. "Every kind of nut out there in the world wants a jet engine."

The 2006s are in: this state-of-the-art kart arrives standard with push-button jet ignition and side rails that double as fuel tanks for gasoline, kerosene, whatever. The builder has reached 60 mph on back roads. Your mileage may vary.

Ersatz Elevator

Space elevator model

MAKER	**Jack Buffington**	COST	$500
OCCUPATION	Contract electronics designer	TIME	10 hours
LOCATION	San Francisco, California	MORE	buffingtonfx.com

Everything Jack Buffington does involves a little sleight of hand. He's held a job making animatronic birds to lure hawks into giant nets. He's done movie special effects. Lately, he makes secret remote devices for stage magicians. Buffington, 31, does all his work in a former factory space in the San Francisco Bay naval shipyards.

One of his most challenging feats involved making a robot climb a ribbon with no strings attached. Buffington heard of a contest from the Institute for Scientific Research to create a mini space elevator, the fabled device that aerospace enthusiasts believe will shuttle visitors beyond our atmosphere along thin strips of carbon.

His final scale model weighed just 2 ounces. He machined the brass gears and shafts himself, and created the structure from thin sheets of folded polycarbonate. Feeling pretty cocky about winning, he bought tiny, high-efficiency solar panels, the same ones used in NASA satellites. During the contest, the panels would receive rays from lights positioned on the ground for a total 2,000 watts of power. It was no surprise that Buffington's model ascended 12 feet of orange safety tape in just over a minute, and won.

As a kid growing up in Pennsylvania, a neighbor set Buffington on course with a spontaneous gift of a RadioShack 101 project kit. He attended three semesters of college at Penn State Behrend ("Mostly general education classes, and mostly flunked out"). Then, after moving to San Francisco, he studied model building at the Academy of Art College.

Was Buffington a space fanatic as kid? "Um, sure," he says — but soon confesses the real reason for his elevator project. "They were offering a $1,000 prize."

The tiny climber — measuring 4 by 4 by 8 inches — ascends a piece of plastic tape using the power of 66 high-efficiency solar panels. Bright electric lights nearby shine on the device, enabling it to reach 12 feet high in just over a minute.

Cool Shirt

Shirt with 38 working fans

MAKER	**Mikey Sklar**	COST	$70
OCCUPATION	Unix system administrator	TIME	10 hours
LOCATION	Brooklyn, New York	MORE	electric-clothing.com

Mikey Sklar, 28, is a vice president in the high-performance computing group at a tony investment bank. But don't let that fool you. He's also an innovator of electric clothing. Take his fan shirt, which uses recycled materials put together in his Brooklyn apartment to keep him cool in the Nevada desert.

To create the shirt, Sklar used an elasticized cord interlaced between the empty bolt holes in the computer fans (the elastic helps the fans move around a bit with the wearer). A spool of wire he found in a dumpster connected all the 12-volt fans in parallel. He runs them off eight rechargeable NiMH AA batteries in a store-bought battery pack. A switch from an old disk storage array turns the fans on and off, and twisties neaten the power cords.

The fan shirt does not exactly qualify as business casual at Sklar's job. "I sort of doubt there are many DIY people on Wall Street," he says. Soon after dropping out of the University of Washington, Sklar moved from Seattle to L.A., where he started working for the investment bank as a 20-year-old. He got the job through a consulting company, but the investment house soon asked him to move to New York, and poached him, paying the consultant a fee to break the contract. Sklar's specialty is "out of band" hardware, networked devices that control and test large mainframe systems remotely.

The road that brought him to his current position was more tied to Sklar's informal hacking than you would think. "I've always had a lot of hobbies and I've let them be dominant in my life," says Sklar. "I was a horrible student." Sklar says he was heavily influenced by both parents: his mother, who immigrated to the U.S. from Panama, taught business computing at the University of Washington; his father met her in South America where he ran the databases that tracked goods flowing through the Panama Canal.

"By the time I got to college, it was clear that the technical world was getting advanced, people were getting wealthy very quickly, there were a lot of jobs out there," he recalls. "I spent a lot more time getting further into Unix at school, to the point where I was working as a sysadmin for AT&T Wireless. My last semester, I failed ethics and went full time with AT&T."

In his spare time, Sklar helps his girlfriend run a large clothing swap party, in which hundreds of people toss their used duds in a pile and everyone gets to pull something out. His sewing projects include a series of conductive pockets that prevent RFID readers from penetrating your clothing to read the products or ID cards you may be carrying ("I'm basically building a pocket for the tinfoil-hat people," he says). There's also an LED-embedded dress he sewed that responds to ambient noise. As the volume in the room increases, waves of light climb higher up the dress.

With the fan shirt, Sklar encountered only a few problems on a test run at the Burning Man festival in Nevada's Black Rock Desert. It really does cool you off, but it's heavy at over five pounds and lasts just three hours. And though a clerk at a concession stand gave him free coffee because she liked his getup, the contraption also drew a lot of pesky questions. "When I wear the shirt I'm often asked, 'Do the fans spin?' Well, of course they do!" says Sklar. "Who would walk around wearing a bunch of computer fans that don't spin?" Indeed.

N U.S.A.
M1107
18ca

Two Wheels Good

Robot that can balance on two wheels

MAKER	**David Anderson**	COST	$1,000
OCCUPATION	Computer laboratory director	TIME	A couple years of fine-tuning
LOCATION	Denton, Texas	MORE	www.geology.smu.edu/~dpa-www/robo/nbot/index.html

If you want to make great robots, you have to keep a close eye on the Saturday morning cartoons. "Cartoon characters don't just start running," explains David Anderson, 53. "If you look carefully, they take a step backward first, then put their arms out in front of them, then run." Anderson explains that cartoon runners, like other bipeds, know to shift their center of gravity in the direction of travel. Ever since he built the first prototype for a robot that could balance itself on two wheels, Anderson has made a vast, obsessive study of the ways humans balance during ordinary activities. In the space of an ordinary conversation, he'll describe how sprinters explode out of starting blocks, how prize-fighters prepare for a blow, and how babies first learn to use stairs.

Anderson, a computer programmer and the director of the Geophysical Imaging Laboratory at Southern Methodist University, has created one of only a handful of self-balancing robots in the world. The bot exhibits amazing poise when zipping up ramps, stopping, turning, and marauding over uneven ground. He built the device in his home workshop as a hobby project and to show off at robotics clubs.

Like the cartoon characters, the balancebot moves backwards an inch before moving forward. Then its center of gravity shifts, and leads the robot into its next direction. All its movements appear a little wobbly, like a toddler with a high center of gravity (in this case, a heavy battery pack). When it charges down an incline, the robot will momentarily speed up to regain its center of gravity ("like a little kid running out of control down a steep hill") and then slow down again. Anderson's mother, a former botanist, was so taken by the machine's pathetic failures to negotiate a rutty Texas road that she gave it a name: Baxter.

While chipping away at the code, Anderson was conducting informal field research the whole time. He takes his brown-bag lunch to work, parking himself outside Heroy Hall on the SMU campus most days, looking for clues among the bicyclists, rollerbladers, and skateboarders. "When a pack of skateboarders finds a new railing or set of stairs, it takes them about 20 minutes to learn the physics of the new surfaces. You see them trying to lift the board on the rail and grind down. The first few times the guy falls right on his ass. But in a short time, he learns how to keep his weight back, and shift it forward at just the right moment to keep his balance."

Robots don't learn like skateboarders, but they do fall. The balance bot's next challenge is finding a way to right itself. Anderson recalls a kung fu master on TV once leaping from a prone position without using his hands. It made a strong impression on him. The roboticist briefly considered a similar move for his machine until he decided the motors probably wouldn't have enough torque. Now, he's thinking about a servo-assisted arm that would push the robot back up.

The exercise of building two-wheeled bots began when his father, a 79-year-old former physics professor, told him about the problem of the inverted pendulum — a mechanical or software system that can keep a heavy weight stabilized on top of a long stick. Right away, Anderson started working on his own inverted pendulum bot in the garage of his suburban three-bedroom ranch. He does all his own metal work, and half the room contains a vertical mini-mill, 8-inch lathe, »

The Journey Bot is another recent project, created with engineer Mike Hamilton. The 20-pound self-guided vehicle avoids obstacles using four sonar detectors in front. Guidance data comes from the GPS receiver on top and an inertial measurement sensor mounted on the pole.

Anderson at school trying to push his bot off balance. The 10-pound, 12-inch robot can recover from all but the most violent shoves; out of doors, it tackles rocky terrain with only a rare fall.

» and various electronics test equipment. (The other half contains a costume-jewelry workstation for his wife, who makes pins and broaches as a hobby and sings professionally with Dallas Opera.)

The first task was to teach the machine which way is down. The balance bot started with an aluminum feeler connected to a potentiometer as a ground sensor. Anderson advises builders to use the crude but robust feeler at first, perfect their software, and then move to more advanced components. The bot now orients itself using a piezoelectric gyroscope and ADXL202 accelerometer run through Wiener and Kalman filters. (The sensors feed into a Motorola-based microcontroller.)

Anderson's advice for software coders? Remember that robot movement is an aesthetic challenge as much as an engineering one. "It's like tuning a piano. The octaves aren't equal distances apart. A person has to bring subjective experience to it," says Anderson, whose first career was as a percussionist. He learned computers as a hobby in the 80s, and he now manages a team that makes geological models of the Earth and other planets. The robot building started years ago when his two boys, now 20 and 22, were going through a Lego period. It seems to have stuck with Dad.

Now, with the practice of dozens of robots under his belt, Anderson says his ultimate goal is to achieve lifelike movement. "Robots can be programmed to have that jerky robot motion or not. It depends on what you're going for," he says. "I want to make one that's as beautiful and fluid as a speed skater."

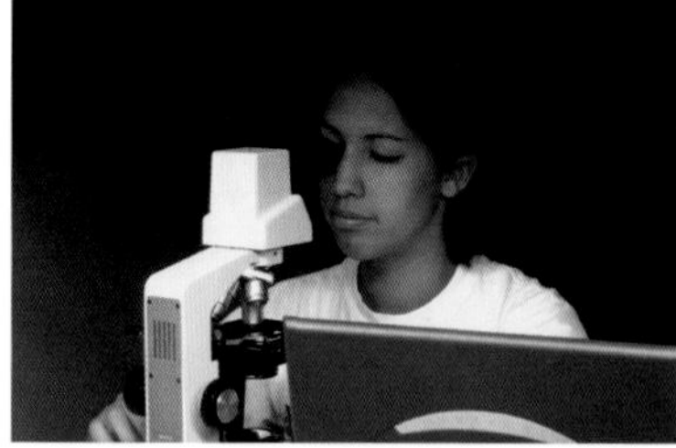

Sarah Mims, 19, in Padre Island, Texas (left), where she's capturing fungal spores carried through the atmosphere from Central America. Her prize-winning research apparatus consisted of a toy store kite, plastic cup, binder clip, and microscope slide.

Kite Aerial Poaching

Dirt-cheap atmospheric analyzer

MAKER	**Sarah Mims**	COST	$25
OCCUPATION	Student	TIME	10 minutes
LOCATION	Seguin, Texas	MORE	earthobservatory.nasa.gov/Study/SmokeSecret/

The National Oceanic and Atmospheric Administration has a budget of $55 million to fly sophisticated air samplers around in aircraft. The U.S. Department of Homeland Security is funding costly samplers of its own in search of airborne biological activity. Sarah Mims, as a high school student in Seguin, Texas, already did it with a plastic cup and a kite. In fact, her low-tech apparatus was so successful that her work was chosen for inclusion in the science journal *Atmospheric Environment*, a serious peer-reviewed publication.

Mims needed the kite to grab spores of bacteria and fungi that she thought were traveling through the air. It was 2003, and she was trying to prove for a school science fair that smoke from fires can carry live spores trapped in carbon particles many thousands of miles. Mims had her parents drive her to a beach on the Gulf Coast. That day, she knew that large fires were burning in the Yucatan. She'd checked NASA weather models to find that the air currents had come directly from Central America. She was ready to assemble her test system right there on the beach.

"I started with a delta-style kite from a store," says Mims, now a biology major at Texas A&M University. "I had a plastic drinking cup and punched holes in the top rim, then tied strings to the holes." She cut a slit in the bottom of the cup, pushed a microscope slide in, and secured the end with a binder clip. She had the routine down pat. In previous experiments, she had smeared Vaseline across the glass slide to make the surface better attract particles in the air. "But things stick to the glass anyway — like a dusty windshield," she says. The kite flew at approximately 100 feet for 30 minutes, then she quickly pulled out the slides and applied clear packing tape to preserve the samples from ground contamination.

Mims and parents raced home to examine the slides. (Her dad, Forrest Mims III, a well-known author of science books, just happened to have an $800 National DC2-156-S digital microscope perfect for the job.) She saw plenty of soot and spores. "Most of them you can see clearly, and I was able to identify one as a kind of *Alternaria*," says Mims. A week later she returned to the beach and repeated the experiment when there were no fires burning on either side of the Gulf. Eureka! No soot, and far fewer spores. "Given the connection between the carbon and spores, I'm pretty confident the spores were from Central America," she says. "But I couldn't have done it without the kite."

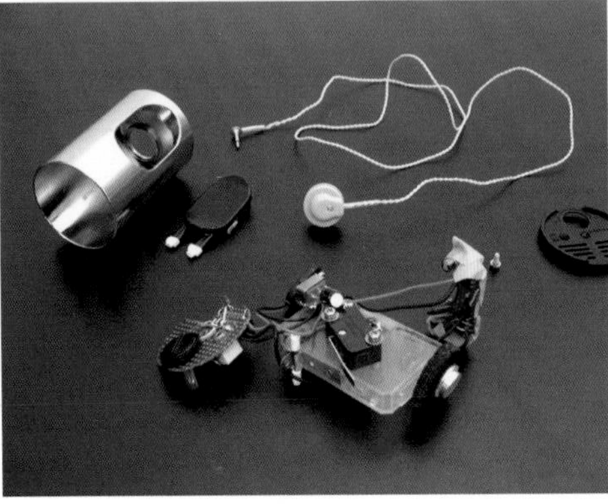

Whenever gray-headed flying foxes make ultrasonic cries, Mitchell (top) hears a tone from his bat detector. He made the device from an old cigarette lighter, inkjet cartridge, and hand-soldered electronics (left).

Batman

Ultrasonic bat detector

MAKER	**Scott Mitchell**	COST	A$30 (about US$23)
OCCUPATION	Post-grad design student	TIME	30 hours
LOCATION	Melbourne, Australia	MORE	hypertext.rmit.edu.au/~smitchell/?p=26

To look at them, the gray-headed flying foxes of Melbourne aren't very cuddly. These bats have a 3-foot wingspan, red fur around their necks, and the ability to fly at 35 mph for hours at a time. Nevertheless, Scott Mitchell, 35, had a very moving experience one night at 2 a.m. on a deserted road as one flew just overhead. And in Melbourne, he notes, flying foxes have been shot by a misguided government program and need to be protected. Mitchell decided to use his hardware modifying skills to communicate with the bats, or at least hear what they had to say.

He resolved to change a beautiful old cigarette lighter — a 1968 Braun Cylindric T2 crafted by design legend Dieter Rams — into a device that would signal when bats were voicing ultrasonic sounds. Some bats use these sounds for navigation, some use them to stun prey, but flying foxes seem to simply talk to each other. Mitchell thought there was something "mesmerizing and beautiful" about their cries.

Not mesmerizing enough, however, to ruin a classic design piece. Mitchell decided to modify the heavy, chrome-plated gadget so it could be repaired with its original parts and brought back to mint condition at any moment. He's working on a second career in industrial design by getting his master's at RMIT University in Melbourne, and he says he enjoyed the challenge.

Mitchell started the project by buying an ultrasonic transducer online for $4, fitting it where the flame should be. A transducer operates like a microphone, changing the ultrasonic frequencies into an electrical signal, which is then amplified by two audio amplifier chips. Loosely following a design he found online, he soldered in a binary counter chip, which divides the frequency by 16, acting as a crude analog-to-digital converter. ("The resulting audio sounds a little like a Geiger counter," he says.) Mitchell used a spent inkjet cartridge to house the circuit.

While getting his master's, he's teaching a class about design to undergrads. Their first assignment? Create a "pokia" phone, a device that uses a vintage phone handset to connect to a mobile phone. He tries to impart to his students that there's more to tinkering than just a fun pastime. "Basically, I think modding is a political act," he says. "Modernism changed our relationship to objects, remaking us all into consumers. But when we hack a product, we suddenly realize that things are not the fixed objects we pretend they are — there's more to them, awaiting escape."

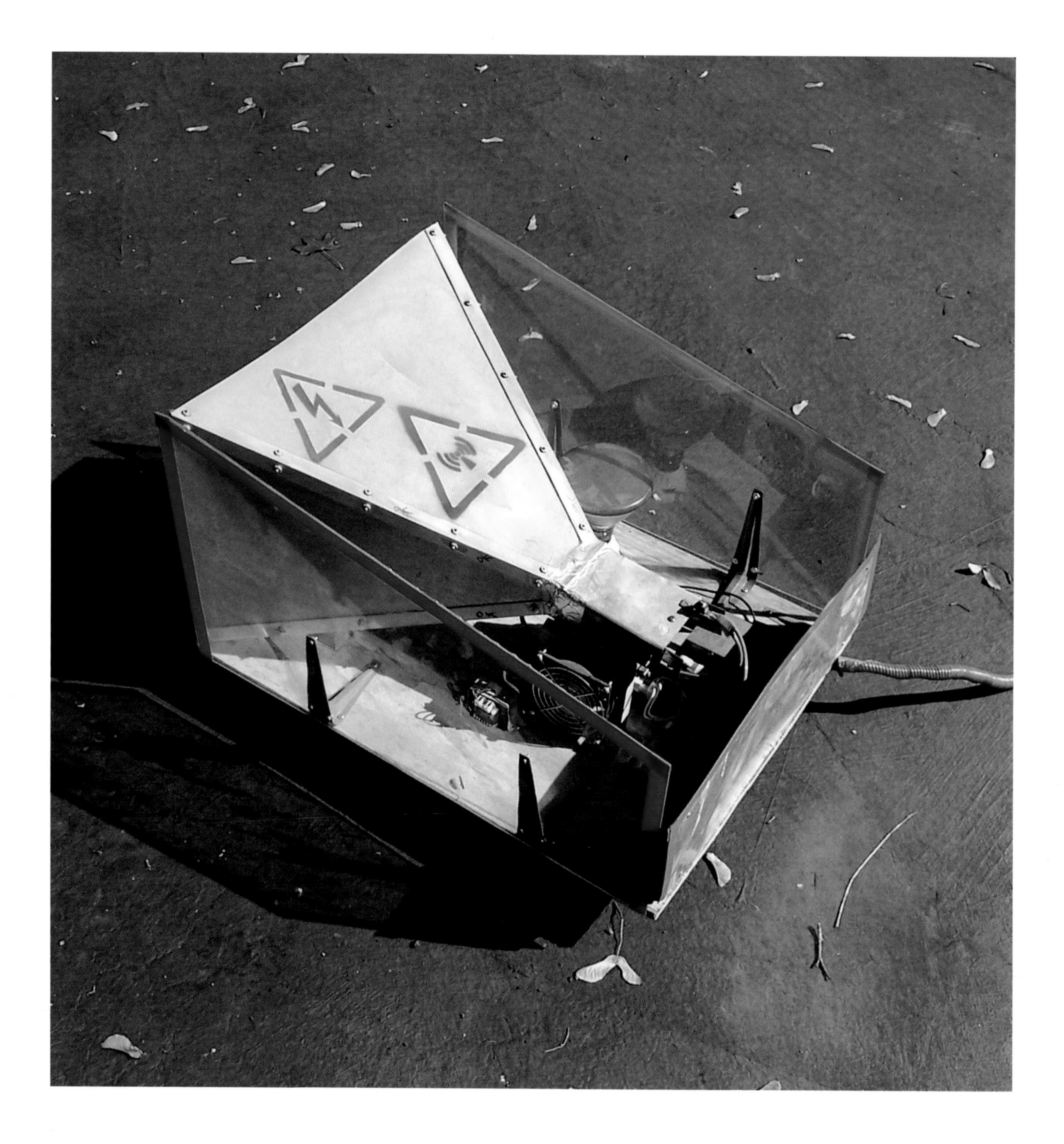

Shoot Out the Lights

High-energy radio frequency gun

MAKER	**Slava Persion**	COST	$600
OCCUPATION	Student	TIME	2 months
LOCATION	Nanuet, New York	MORE	voltsamps.com

It's amazing what kids will get up to when they're bored. A few years ago, Slava Persion, 22, decided to make a HERF gun during his summer vacation. Rhyming with "Nerf," HERF stands for High Energy Radio Frequency, and it's a device that has the ability to short-circuit electronic equipment wirelessly at 20 paces. Once you aim the antenna and fire, innocent household communications equipment doesn't stand a chance.

Once he fixed on the idea, Persion, who's now majoring in computer science at New York Institute of Technology, combed his quiet residential neighborhood for an old microwave during bulk trash day. He gutted it for the power supply and magnetron, a component that oscillates high-voltage electricity. The final touch was a handmade box-shaped antenna that he created from scrap steel.

After a little fiddling, he soon had the HERF running in his bedroom. "I noticed my computer screen was flickering when I brought it close," recalls Persion. "I realized there was a connection there. I was really excited."

Persion says he became "power hungry" for a much bigger model. On a whim, he took a summer job at Kohl's department store selling shoes to save up for the components he would need on eBay. And he began taking his physics homework a lot more seriously. To make HERF 2.0, Persion found a 50-kilowatt magnetron online for $300. It was from a military radar system, and would have more than 50 times the power of his puny microwave hack. He matched it with a neon-sign transformer that he'd found in a dumpster.

The results didn't disappoint. This super-HERF crashed computers throughout the house and shorted out the cordless phones. "It was the coolest thing I'd ever seen," he says. "But then it sort of self-destructed. The tube started arcing. There was a mess of high voltage everywhere. The whole chaotic scene was over in a few minutes."

When Persion booted his father's PC back on, it appeared to be DOA. "My dad's computer had a black screen, so I was like, 'Oh, no, this is bad,'" says Persion. After several attempts, the computer started again, and Persion pretended like nothing had happened.

The child of Ukrainian immigrants, Persion came to America when he was eight. He claims his father, a computer programmer, and his mother, a piano teacher — both of whom he still lives with — are only vaguely aware of his exploits. "When I showed her the HERF, all my mom did was kind of smile." Once Persion put his world domination machine up on his homepage, he got attention. He appeared in a national business magazine and earned an invitation to demo the machine at a military technology conference.

For all his aptitude for improvised radio weaponry, Persion describes his academic record as rather mediocre. "School's not on my side in terms of learning," he says. "I'm what they call ADD, which is a personality trait, really."

Persion claims he's currently working on a (legal) product that would change public traffic lights for ordinary citizens. But he's also been building better HERF guns using the full metal shop he's pulled together in his garage over the last few years. His second device self-destructed into a pile of smoldering parts because the circuit missed a crucial diode. The correct parts were on the way to him in the mail when he fired it up. "The problem was I just couldn't wait," he says. "I'm impulsive."

The couple's Ford truck, above, looks antique thanks to the blueprint camera. The 3-inch-square homemade device (left) was constructed from a champagne box, an old lens, and weather-stripping rope to keep light out.

Tangled Up in Blue

Camera that uses blueprint paper instead of film

MAKER	**Liz Zazulak and Reagen Ward**	COST	$10
OCCUPATION	Homemaker, Unix systems administrator	TIME	3 hours
LOCATION	Austin, Texas	MORE	blueprintphotography.com

Talk about a cheap thrill. When Liz Zazulak and her husband Reagen Ward taped some architectural blueprint paper inside a broken Polaroid land camera, they weren't sure what would turn out. Blueprint photography hadn't been described in any book, and even on the net, only a few people admitted to trying it. Early signs were encouraging, says Zazulak, 32. She peeked into the camera a few times over the three-hour exposure: "The way the colors were shifting was just fantastic. With the paper, blues become white, and oranges and yellows stay much darker than they should."

Soon they had an "amazing" photo. The subject was simply their drab backyard featuring a satellite dish — "I called it *Trashy Suburbia,*" jokes Ward, 32 — but the process was like a new frontier to both of them. Zazulak, a serious photographer who took a break from her master's degree to have her daughter Bailey, 2, was more interested in the artistic potential. Ward, a Unix administrator for payroll company ADP, was more caught up in the technical details.

Blueprint paper costs around $30 for 300 huge 2-by-3-foot sheets. Ward scored some by visiting a large reprographic shop, hat in hand, asking to buy a stack for an "art project." The material is covered in diazonium salt crystals, which break down in light, and then can be fixed by a bath in ammonia. Unlike photography paper, blueprints create a positive image, so you instantly see what you get. Zazulak and Ward were so thrilled, they started a website and created several home-built cameras especially for the new media.

"This blurs large-format photography with alternative processing, with people who are just bored," says Ward. For this reason, inexpensive lenses are usually easy to come by from people who are just interested in helping out. Probably the Zazulak and Ward's most stylish home-built camera is one made from a champagne box, which uses a scratched photographic lens bought online, for $7 total. (Lenses also come from discarded overhead projectors and other office detritus.)

Ward drilled a large hole in the long narrow box. Inside, he pressed sticky weather stripping along the joints as a light seal. Using thumbtacks to hold it in place, he attached the lens. Zazulak created a focusing screen — a little plate in the back of the camera to adjust the blueprint's distance from the lens — by taping wax paper inside and marking the lens' focal length with a pencil. She cut a piece of blueprint paper to fit the dimensions of the box (3 by 3 inches), and let it go.

As the material soaks up light inside the camera, an image emerges over several hours. Zazulak and Ward pulled out the blue sheet from the champagne box, and it was a fully realized image of their old truck, a favorite subject because the paper makes the vehicle look antiqued and charming. To fix the print, Ward zaps a few tablespoons of regular laundry ammonia in a microwave for a few seconds. He pours the steaming liquid into the bottom of a bucket, tapes the blueprint inside, and seals the top with a lid.

Even with the fixing, the diazonium crystals fade. While normal photography provides a permanent record, these images stick around only a few years. Zazulak and Ward's famous first *Trashy Suburbia* image, for example, has almost disappeared.

With thousands of volts blasting the fusor's core, the purple glow signals that hydrogen is entering a plasma state (top). The image was shot with a digital camera sensor inside the stainless steel housing. Left, Hull stands between his power supply and fusor.

A Real Mashup

Homemade nuclear fusor

MAKER	**Richard Hull**	COST	$3,000
OCCUPATION	Electrical engineer	TIME	1 year
LOCATION	Richmond, Virginia	MORE	fusor.net

Like many Americans, Richard Hull is doing his part to solve the energy crisis. Some people inflate their tires for better mileage, some caulk their windows and doors against drafts. Hull, 59, built a thermonuclear device.

It's serious science, but not as unusual as you'd think. Hull is doing fusion, the act of combining two hydrogen atoms to make one helium atom. It could be the key to releasing a huge amount of energy, and over the last decade, nearly two dozen amateurs have created small nuclear fusors in garages and basements. The method they use was developed by the inventor of the television, Philo T. Farnsworth, in the 1960s. It's called inertial electrostatic confinement; the fusor bombards hydrogen atoms with very high voltages until the particles accelerate and collide.

"Oh, fusing atoms is easy," claims Hull with a slight Virginia twang. Well, OK, not that easy. First you have to find a way to generate over 20,000 volts of electricity. Hull refurbished the power supply from a discarded hospital X-ray machine. He bought two 10-inch stainless steel hemispheres from a machine shop — basically heavy-duty salad bowls which contain the reaction. Parts were machined in a 600-square-foot home workshop boasting lathes, milling machines, and oxyacetylene, MIG, TIG, and spot welders.

To start the fusor, Hull pumps all the air out of his hermetically sealed salad bowls with a high-powered vacuum pump. He introduces gas containing a hydrogen isotope into the chamber (half a pint costs $55 online), then looks through a small CCD camera inside the steel to monitor the reaction and capture images.

Fusor enthusiasts live for the brilliant purple light that emanates from the reaction — an indication that the gas has passed into a plasma state and that neutron radiation is on its way. The whole project can be achieved for under $1,000, though Hull has splurged on a neutron counter costing nearly that much to verify that he's reaping tiny amounts of energy during tests. He's quick to point out the dangers of the procedure: "With this project, you can be killed in so, so many ways! Electrocution, gas explosion, flying metal shards, irradiation with X-rays and neutrons."

But science needs a little peril to make it engaging and vital, believes Hull. "When I was 12, it was a different world," he recalls. "Sputnik had just gone up and the government was desperately trying to train future rocket scientists. Twice a year the government turned over an army base to rocketeers. We mixed hundreds of pounds of fuel and fired it in solid-fuel missiles. The rule was: you put sandbags around your work area. If you blew up, it would kill only you."

An electrical engineer by day, Hull organizes meetings of the High Energy Amateur Science club in his spare time. The 80-person gathering convenes once a month at his house where they compare fusors, Tesla coils, and devices that create water arc explosions. After going around the block a few times with his fusors, Hull has his doubts about whether anyone could produce a surplus of energy. All of his models require much more energy than they reap. All the same, he enjoys a challenge that combines diverse fields such as materials science, vacuum physics, and high-voltage electricity. "We're actually doing what the distant stars do," he says. "There is some possibility that you could uncover a core process of the universe. Nothing's more enticing."

Blinding Technology

Halloween costume with 24 camera flashes

MAKER	**Rob Cockerham**	COST	$108
OCCUPATION	Pre-press operator	TIME	50 hours
LOCATION	Sacramento, California	MORE	cockeyed.com

Talking with 36-year-old Rob Cockerham, one soon realizes that most of his energy is focused outside of work. A pre-press technician by day, Cockerham runs a personal website that attracts hundreds of thousands of monthly viewers. He stages gonzo science experiments that draw notice of the national press. Even the site's lab assistants are beautiful and inexplicably numerous.

But the Sacramento, Calif., native is perhaps best known for his Halloween costumes. He's gone as a hideous fish man, the wooden game known as Jenga, and as a papier-mâché continent of Africa. His most technically challenging costume had him rewiring 24 one-time-use cameras so that they would all flash around him on cue, a piece he calls *Paparazzi.* "This was probably the peak of my soldering career," he says.

Besides some very basic home electronics, the costume made the most of scrounged materials Cockerham found around his neighborhood. First, he knew you could bridge two contacts on the circuit board of a single-use film camera to make the flash go off, so he drove around to the photo departments of Target and Rite Aid Pharmacy, telling the local clerks about his planned Halloween getup. He soon had more than needed. At RadioShack, he bought two 12-position rotary switches for a home stereo and 100 feet of cheap telephone wire. A friend helped him locate the contacts on the circuit board to discharge the 300-volt capacitor, thus triggering the shutter. He soldered one contact to a common pole on the RadioShack switch and another to each individual position on the switches.

To make the cameras look real, he spray-painted the tops of 2-liter soda cans he found on the street, and mounted the setup on a sheet of his favorite material, a corrugated plastic called Coroplast, which he gathers from spam political signs and weight-loss advertisements nailed to the telephone poles in his neighborhood. "That's the best stuff," says Cockerham. "It's light and strong. This piece says, 'Robbie Waters — For All the People' on the back."

On his one and only night using the costume, Cockerham won first prize at the Lucasfilm company Halloween party in San Rafael, Calif. It was his revenge against George Lucas for winning the measly "door prize" at the party the year before, and confirmation that a guy from Sacramento could, ahem, "strike back" at all those professional model-makers.

In creating all his projects, Cockerham has a strong work ethic about the process of making things. (His dad, a forensic scientist for the California justice department, always encouraged tinkering at home.) "It's OK to take a long time putting things together," he explains. "First, with a real, physical project you can't make mistakes — there's no Last Save button. But there's a little bit of joy once every little step is complete. Then you realize it really might work."

Shocking Developments

Washtub Tesla coil

MAKER	**Matthew Stiger**	COST	$300
OCCUPATION	Electrician	TIME	3 days
LOCATION	Seattle, Washington	MORE	flaminggasneon.com

Matthew Stiger, an apprentice electrician in Seattle, sure suffers for his hobby: shocks, burns, even head trauma. On his latest Tesla coil — a 7-footer — he built a beautiful dome from two 12-inch Ikea salad bowls supported by an aluminum cylinder. When it toppled over, he got a forehead gash that required 11 stitches.

Why all this trouble for a device that's useless except for its ability to shoot really cool sparks in the air? Serious electronics fans consider them the benchmarks for basic proficiency. Tesla coils are like triple layer cakes for bakers, or double axel jumps for ice skaters. Though Stiger, 38, counts himself as a mere intermediate "coiler," he won the 2005 Dorkbot Teslathon, a juried competition among coil builders in Seattle, a big honor considering his unorthodox design.

Stiger's winning coil featured an old, rusted washtub atop the various coils of wire, a dramatic switch from the shiny, doughnut-shaped "toroids" atop most designs. ("Toroids are tired," he declares.) The washtub turned out to be large enough to store a significant surface charge, making unusually long, 7-foot sparks for a relatively modest 14,000-volt coil. It also produced wide, flat arcs along its rim, which were a hit with the Dorkbot crowd. "The sound really is thunderous, and the tiny room became warm fast, and full of the smell of ozone," Stiger recalls. "It was freaking everyone out. They loved it."

It's not exactly as Nikola Tesla intended. The 19th-century inventor hoped the coils would one day transfer household electricity without wires. Stiger is not sure why he likes them, just that he always has. "Electric sparks seem to be moving around doing their own thing, it's like they're alive. My mom tells me that whenever we had electrical storms in Hammond, Indiana, my sister and brother would run and hide, but I'd want to watch the whole thing."

His stepfather described a Tesla coil to him when he was 12, and he immediately tried to put one together with some wire and a toilet-paper roll. (It didn't work.) Since then, Stiger's built more than 30 coils. He works on projects out of his Seattle warehouse district apartment, which he shares with two cats, a homemade lathe, and thousands of castoff industrial transformers and capacitors. He says it has few luxuries, except for the 200-amp service panel in back.

For the washtub coil, Stiger built a power supply from a 14,000-volt transformer bought for $15 from a Seattle City Light salvage yard. It featured a homemade slide-cord choke, a device that keeps the high voltage from returning to the wall. Stiger hand-wound the inductor coils around a cardboard tube he found in the trash near a carpet store.

Stiger's romance with electricity has always been a factor in his career choice. After a few semesters of community college and Army service as a linguist (Spanish), he completed trade school to build neon signs, a trade he still practices part time. Two years ago, he began looking for a higher-paying job. Anything like sales was out. ("If I went out and tried to sell something it probably wouldn't work," he laughs. "I'm kind of a goofy character. That may be why I'm still single.") Instead, he's been training as an electrician, working on the site of a new Fred Meyer department store during the third shift, installing junction boxes, conduit, and transformers. The icing on the cake? Great health insurance.

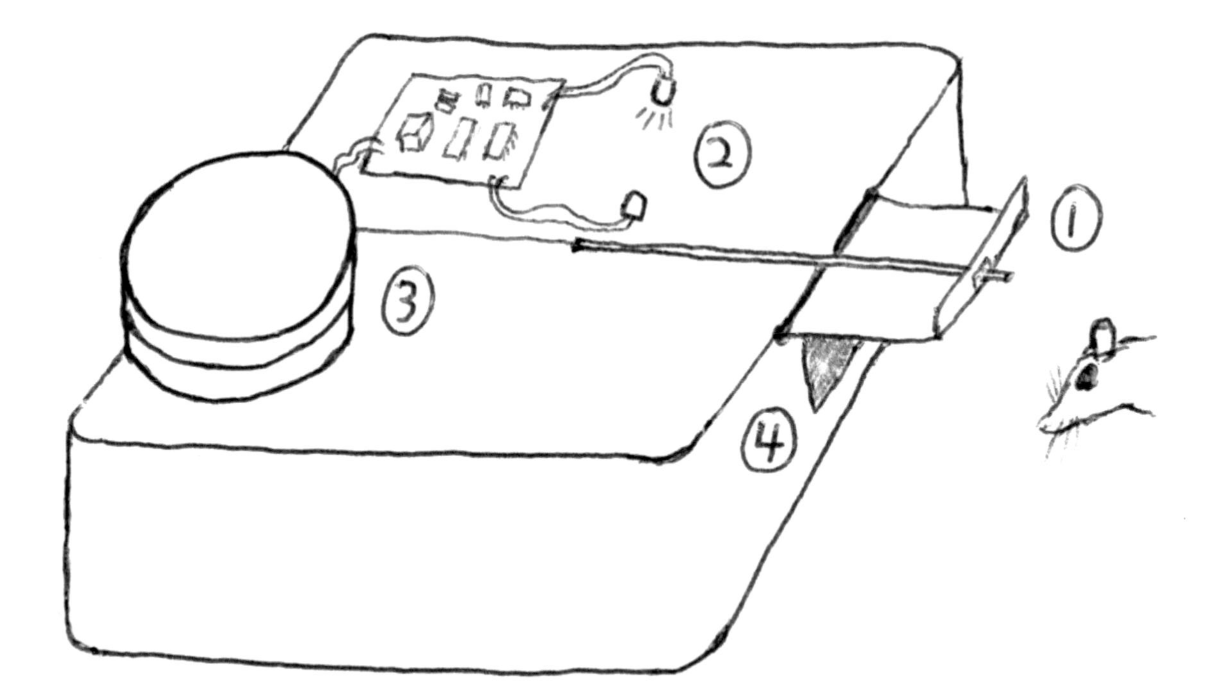

1. Mouse enters trap
2. IR beam broken near bait
3. Motor retracts pin
4. Door falls shut and seals

Catch and Release 1.0

Mousetrap hacked from junk

MAKER	**Neil Fraser**	COST	Can$0.10 (US$0.08)
OCCUPATION	Web programmer	TIME	4 hours
LOCATION	Elgin, Scotland	MORE	neil.fraser.name

The mice visiting Neil Fraser's high-tech rodent hotel have no idea how lucky they are. Not only do they witness a state-of-the-art infrared detection system, but they get a free ride down the street from Fraser. The owner of a Scottish web programming company takes them a few blocks away to a woody park and gently shakes them loose from an old shortbread tin.

Fraser caught three mice in his trap immediately after its installation. To make the device, he scavenged a 5.25-inch disk drive, installing its infrared LED and photo transistor at the end of a toilet paper tube. When a mouse crawls down the tube, it breaks the infrared beam, causing the floppy drive's motor to spin. This retracts a coat-hanger wire holding the trap door open. The door, with hinges made from a clothes drier part and a drinking straw, snaps shut thanks to five fridge magnets. An old vending machine switch detects that the door is shut, and cuts the power to the motor.

We have only blurry pictures and some blueprints as this trap's legacy. One of the inventor's relatives accidentally threw the trap away, not unreasonably thinking it was a piece of trash. It's true that all the materials came from junk in Fraser's workshop and his adventures dumpster diving around his native Ottawa, Ontario. A few years ago, he moved to Elgin after getting a programming job at a Scottish company. Once he established himself, he started his own company.

Fraser, who majored in cognitive science at Carleton University, yearns to build a better mousetrap — perhaps one that uses a microcontroller — but he's been stymied by living in his adopted country. "The big problem is that in Elgin there's nowhere to buy parts," he laments. "If I need a 10K resistor, I have to mail order it, which means I pay a lot for postage and I still might have the wrong part. It's impossible to do proper tinkering in this town. There's no RadioShack, there's nothing."

Back when he was living in Canada, he made quite a few strange contraptions, including a computerized Etch A Sketch. The hack allows you to transfer any digital picture onto the classic toy via a PC. A set of stepper motors turns the knobs on the Etch A Sketch using Meccano parts. The device employs motors from an old fax machine and tape recorder, and a power supply from a Wang mainframe. "I guess I miss the dumpster diving the most," he says.

GENERAL ELECTRIC
SYNTHESIZED
PROPERTY OF DAVID FORBES

Retro Chic

Nixie-display wristwatch

MAKER	**David Forbes**	COST	Scope clock: $2,000; Nixie wristwatch: $10,000
OCCUPATION	Electrical engineer	TIME	Scope clock: 1 year; Nixie wristwatch: 3 years
LOCATION	Tucson, Arizona	MORE	cathodecorner.com

For David Forbes, tinkering turned out to be a way to cope with a very difficult situation. In mid-2001, his toddler son Linus was diagnosed with leukemia (he's since been treated and has fully recovered). At the time, Forbes was working 50 hours a week for a government contractor, making some of the most rugged, powerful computers in the world. "The job was getting steadily more Dilbert-like," explains Forbes. "Then my kid got sick and I decided I would bow out of the military-industrial complex and do something more sane with my life."

Forbes, 44, quit the job. He created a clock made out of an oscilloscope that received a favorable write-up in *IEEE Spectrum* magazine and began selling wildly to geeks on eBay. He moved on to a wristwatch that used the old-style digits from a 1960s voltmeter. It became a soothing distraction during the height of the family's ordeal. "Everything was all kind of tied together," he says. "I was able to redraw my schematics for the watch while I was at the clinic and Linus was getting filled up with various chemicals. When someone in your family gets cancer, it's a life-changing thing. I think it ended up changing my life for the better.

"I decided to become a full-time tinkerer, building clocks for a living. For me, it harkens to the age of hands-on electronics that I remember my father doing years ago. Nixies are sort of the perfect DIY electronics project."

Forbes has a particular love for Nixie displays. Sold until 1973, the displays were the only means to create electronic digits and letters before the advent of LEDs. When the vacuum tubes are charged with 170 volts of current, they shine numbers and letters with a solid, mellow orange hue. "Nixies produce a display that's fully formed, not pixelated," Forbes says. "There are actual pieces of metal inside making the numbers."

Nixies, however, draw a lot of current. The Nixie wristwatch would need to be highly efficient to work for very long on someone's arm. First, Forbes programmed the PIC microcontroller at the heart of the watch to turn the display on and off quickly so that the user wouldn't notice. Then he devised a scheme in which the display stayed completely dark until it was flipped toward the user. Forbes found a small tilt sensor to fit in the case — the ADXL202 from Analog Devices. Now when you turn your wrist quickly, the watch comes alive and cycles through the hours, minutes, and seconds. The common camera batteries last about four months at 50 viewings per day. At 2.2 inches in diameter and 0.8 inches thick, the 2-ounce wearable is highly accurate, water-resistant, and (some would say) fashionable.

Over the last few years, Forbes has sold a few hundred objets d'art online, but he now realizes it's only a sideline. "One can't make a living as a full-time tinkerer. As a hobby, it pays for itself, but it certainly doesn't pay the rent." Forbes holds a job half time at the University of Arizona's Steward Observatory Radio Astronomy Laboratory, creating spectrometers for the school's telescopes. It's his job to bring the devices' high-frequency operating signals of 200 gigahertz down to 1 megahertz so they can be analyzed by the school's computers.

Ironically, Arizona is the same university Forbes dropped out of in his sophomore year. As a child, he'd done electronics projects with his dad (also a telescope engineer). He'd built an oscilloscope at the age of 12, »

Forbes' other recherché timepieces include a full-sized Nixie tube-based desk clock (bottom right) and a 5-by-5.5-by-10-inch oscilloscope-based clock (top and bottom left). The builder currently has a prototype oscilloscope pocket watch.

The wristwatch (right) with its cover off. The device uses old-school Nixie displays and a gesture-based on/off switch.

» and a Motorola 6800-based computer at 17 with his brother. When he entered college to study computer science, he discovered that a work-study job building a computer-controlled mirror-polishing machine was more satisfying than his classes. "They were teaching stuff at a level so far below what I was figuring out on my own that I didn't see any point of continuing," he says. "Now I work with a bunch of people with advanced degrees in electronics. They're handy to ask about the theory behind things, but I don't know that I missed an awful lot."

When not at the telescope, or cycling around Tucson during the winter in his beloved Batman sweater, Forbes tinkers in a capacious 500-square-foot workshop in back of the house, so the kids can hang out while dad's on duty. (His wife, Terry, recently quit her longtime career as a librarian to start a cancer moms advocacy group, Families Against Cancer & Toxics). Forbes is currently working on a oscilloscope pocket watch using a tiny battery-powered CRT. His son Linus (named after a certain open software guru) is now active in soccer and school activities, and he joins his older brother Henry at the workshop when not otherwise occupied. "It's basically your standard mad scientist's laboratory — with a couch and TV for the kids."

Forbes has an old HP frequency counter and other vintage instruments in his workspace, and of course, everything has old-style Nixie displays. He just likes the way they look. "They're exotic and big and hopelessly impractical by modern standards."

Fetch, Rover

Low-budget household robot

MAKER	**Scott Metoyer**	COST	$250
OCCUPATION	Programmer	TIME	3 months
LOCATION	Corona, California	MORE	potchky.com

The kids in Scott Metoyer's apartment building may not know his name, but they certainly know Pedro. The 30-pound rover paces up and down the halls of the complex, wagging its camera and moving to and fro in greeting.

"He zips all over, trips everybody out," says Metoyer. "I've had a whole group of kids trying to psych it out. Man, they thought it was pretty much the funniest thing, something out of *The Matrix*."

Metoyer, 27, created Pedro in his spare time in his kitchen, naming it after the sidekick in the 2004 film *Napoleon Dynamite*. The rover is equipped with a cast-off laptop, the guts of a Logitech webcam, and a Wi-Fi link so that it can be controlled up to 150 feet away, sending back images to a wireless-enabled computer in his bedroom. You control Pedro using buttons for left, right, front, and back on a 3-inch inset desktop screen. Slider controls on the interface position Pedro's camera, and control the speed of the camera movements in real time. Metoyer created a simple interface in 800 lines of Visual C++, drawing from an associate's degree in programming he recently received.

"Right now he's a rover because he obeys commands on the client software," explains Metoyer. "But I'm working on some autonomous functions, including sensors and a robotic arm I'm hoping to add next."

Metoyer, who does contract programming and works part time at Staples, started with an old Pentium II 220MHz laptop languishing in back of his closet. To create Pedro's aluminum frame and acrylic top, Metoyer used tools at the workplace of his father, a construction site superintendent. He cut the materials to make the structure with saws, drill presses, and grinders, and machined a 2-inch extension for the dolly wheels so they would clear the frame, then bolted them to the motor shafts.

Intentionally challenging himself with a tight budget, the roboticist learned to economize. The rover moves thanks to two high-torque windshield wiper motors from a Saturn sedan. The motors, which are soldered to relays in an H-bridge design, steer the bot like a tank. (With one motor moving forward and another in reverse, Pedro can turn on a dime.) A parallel-port relay board and serial-port servo controller arrived as kits from online stores. He soldered in the guts of an old throwaway Logitech webcam to provide Pedro's eyes, and attached two servos salvaged from the steering of a gas-powered RC car to make the camera pan and tilt.

Metoyer says his electronics workbench is not very glamorous, just a kitchen floor which must be cleared each time he quits: "A crazy girlfriend yells and screams at me if I leave a mess behind ... um, on second thought, don't print that."

83

Super Scope

8-inch Dobson telescope

MAKER	**Katie Barmazel, Ivy White, and Sarah Davis**	COST	$1,000
OCCUPATION	High school students	TIME	280 hours
LOCATION	Santa Rosa, California		

With their arms aching, they took turns on a flimsy fold-up lawn chair and talked each other through grueling work sessions. Sometimes they put on French tapes for a source of distraction. Other times they blasted The Killers, Franz Ferdinand, and Chumbawamba in the dark garage. These three friends — all 16 years old — did everything they could to get through the more than 90 hours it took to make a hand-ground telescope mirror. The result? A powerful Dobson instrument that has won awards and given them a peek at solar systems light years away.

Most people with "homemade" scopes actually order the optics out of a catalog. Katie Barmazel, Sarah Davis, and Ivy White wanted to start from scratch as part of a Girl Scout project. "We had no idea what we were getting into. It all kind of snowballed," notes Ivy, who often acts as the ringleader. The idea started when Sarah's father got an expensive motorized scope from the Discovery Store. One night, in the backyard, the friends marveled over craters on the moon. "There was a thought in my head that, 'Hey, I'd really like to have one of these,'" Ivy says.

Soon, they were spending four hours every Friday night with their scout leader Janet Davis (Sarah's mom) driving back and forth to instructional workshops at the Chabot Space and Science Center in Oakland. Various adults helped with supplies and expertise. They started with an 8-inch Pyrex blank, a flat glass disk, which gradually took on a concave shape after hours of hand pressure with various commercial grits and a cement paver. "You go back and forth in a 'W' pattern over the mirror, then wash the grit off after ten minutes. Then you move the mirror a little and start again. It's a repetitious, very slow process," says Sarah. It got worse when they found out that grinding needed to take place under contaminant-free conditions. Suddenly, snacks were banned from the garage to prevent crumbs from marring the glass.

For the main tube, they bought a 50-inch cardboard concrete form at a local home outlet, and covered it with blue contact paper to hide the brand. To make the base, the girls cut plywood and nailed it together to make three boxes. In a Dobson mount, the various wooden boxes in the base swivel to aim the scope. During construction, the trio of astronomers needed a flat object to act as a low-friction bearing between the boxes. They decided that an old record would work well, pressed against a Teflon sheet, and looted one of Sarah's dad's old discs, which happened to be Rod Stewart's "Some Guys Have All the Luck." Much later, Katie actually noticed the printed label in the center of the record. "When we looked at the name, we said, 'Hey, that's not true!'" she says.

Ivy says a true realization of their feat came home to them on a warm April night when an adult advisor pointed the scope and went inside. "We were looking at the M-Class nebulas, and it was so amazing and so beautiful," says Ivy. On other occasions, they delighted in vivid images of the rings of Saturn and the moons of Jupiter. Then in July, the scope took the Merit and Astronomer's Choice awards at the Riverside Telescope Makers Conference, winning over 600 other entries.

"You know what's nice, though?" Ivy says. "Our telescope is actually much better than Sarah's dad's. It's light enough to take on camping trips. It has a clearer image and we can see even farther."

Ryan O'Horo runs dashboard cables for his custom car PC (top). He says a Chevy Malibu makes an ideal hacking platform. Mounted in the spacious glove compartment are the motherboard, power supply, and XM radio receiver (left).

All Mod Cons

Car with touch-screen PC, Nintendo emulator, and chemical toilet

MAKER	**Ryan O'Horo**	COST	$2,000
OCCUPATION	Systems administrator, assistant photographer	TIME	5 weekends
LOCATION	Dunedin, Florida	MORE	aydiosmio.net

Ryan O'Horo, 21, moonlights as a photographer's assistant on the weekends, and he's found that these gigs involve a great deal of waiting around. His employer takes images of kids' sports teams, so the sessions are typically staged near soccer fields miles from nowhere. Luckily, he does have his 2001 Chevy Malibu.

Using parts he bought over the internet and at local stores, O'Horo decked out his car with a touch-screen PC, GPS guidance system, XM radio, and Nintendo game emulator (with retro game pad, of course). Hear the call of nature? Simply assemble a black fabric pop-up screen over the trunk to access a Coleman chemical toilet.

To create the craft, O'Horo drew on plenty of hardware skills (he works during the week as the sysadmin at a hot tub manufacturer). He started with a 1GHz chip on a VIA EPIA M10000 Mini-ITX motherboard and a Lilliput 7-inch touch screen. He modded the screen so that the PC's Up and Down controls are actually incorporated in the bezel, soldering a small mouse circuit inside the screen. He says he's proud to have incorporated a DC-to-DC power supply that makes the most of the juice in the 12-volt spare battery in the trunk. The machine is stable ("I don't have to pull over to Control-Alt-Delete"), but one trouble spot has been the master interface; for now he has to click through individual windows to change applications.

Still, O'Horo couldn't live without his touch-screen Malibu for all the good it's done him inside and outside of work. "I take my girlfriend to Macy's and let her walk around while I play Nintendo," he says. The whole PC system costs him around $1,600, and O'Horo gained a great respect for the Malibu as a development platform. "Everything in the dash is just a shell. All the cable runs are clear. There's even a little port on top of the glove compartment where you can draw USB and power cables."

The weekend freelancer created the sleek black toilet stall by riveting button snaps onto the trunk, and building a form with copper pipe and brackets. (In it, O'Horo looks like he's processing photos in a makeshift darkroom.)

To other makers out there, his only advice is not to overlook the virtues of a good fabric store. "You kidding me?" he says. "Jo-Ann Fabrics is the best for all those things you won't find at Home Depot or CompUSA."

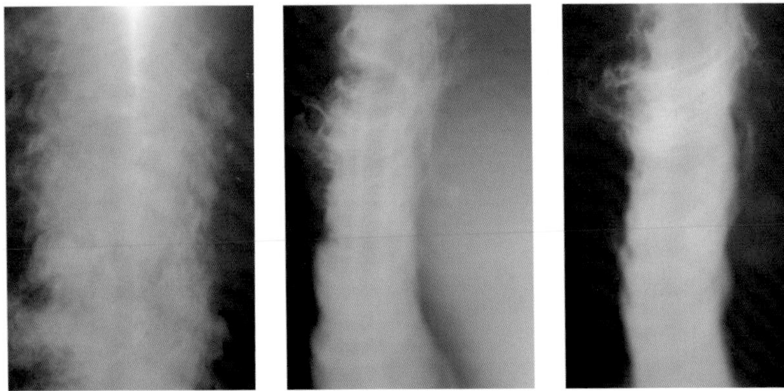

The tornado machine's heart is a mist generator (top). The 5-foot-high cabinet swirls vapor into mini twisters (left) using PVC pipe, plexiglass, and wood.

Surrender Dorothy

Bedroom tornado machine

MAKER	**Harald Edens**	COST	$150
OCCUPATION	Graduate student	TIME	1 day
LOCATION	Socorro, New Mexico	MORE	makezine.com/go/tornado

He moved to the U.S. to see them. He created a huge device in his bedroom to recreate them indoors. They're tornadoes, and Harald Edens, 27, loves them and every other extreme weather pattern. A Dutch national, Edens is getting his Ph.D. in atmospheric physics, studying lightning patterns in three dimensions. When he's not at school or filling his website with extreme close-ups of weather, he's leading trips of Dutch friends who come to the American Southwest to chase after storms in a rented van.

"You do have better weather," Edens sighs. "This is the only place I know of where you can be sure to see tornadoes in May or June. The mountains and spring-time moisture from the Gulf of Mexico create severe storms." When Edens was 12 years old, he read a book on weather and grew interested in capturing images of lightning. His father, a window dresser for retail stores, was a great photographer, and taught him how to capture the best weather candy on film.

Having seen a machine that would reproduce tornadoes at a museum, Edens decided he could make one. He drilled hundreds of holes an inch apart down a length of PVC pipe, creating a system which would conduct air in a circle. Then he cut the 1-by-2-inch wooden supports with a hand saw. A 1-millimeter plexi-glass sheet forms the outside of the machine, which he spray-painted matte black inside so he could see the vortex better.

The finished product was something like a phone booth, with a powerful blower fan from a swamp cooler (eBay, $100). On the bottom, Edens rigged an ultrasonic humidifier and mist generator (hobby store, $30). If you position the misting device in the center, it creates a straightforward swirling vortex up to the top. Move it to the side, and delicate layers ascend.

Of course the real tornadoes are quite a bit more dangerous, and Edens and friends have only come within a mile of them in the Texas Panhandle. He's come a lot closer to nature's fury where he studies lightning, at a mountaintop research facility for the New Mexico Institute of Mining and Technology. His closest brush there came when lightning struck 10 feet away, though a special metal observation tower protected him.

His girlfriend Elisa, from Brazil, isn't crazy about the tornado machine ("The only space I have to set it up is in the bedroom," says Edens. "It's like five feet high, two feet in diameter. She saw the contraption and was like, 'What the hell is that?'"), but she likes following Edens on trips to exotic North American climes. And she did warm up to the idea when it was finally ready to demo. "When it gets working and you get this vortex of mist, it's so cool to watch," says Edens. "Everyone who sees it is in awe."

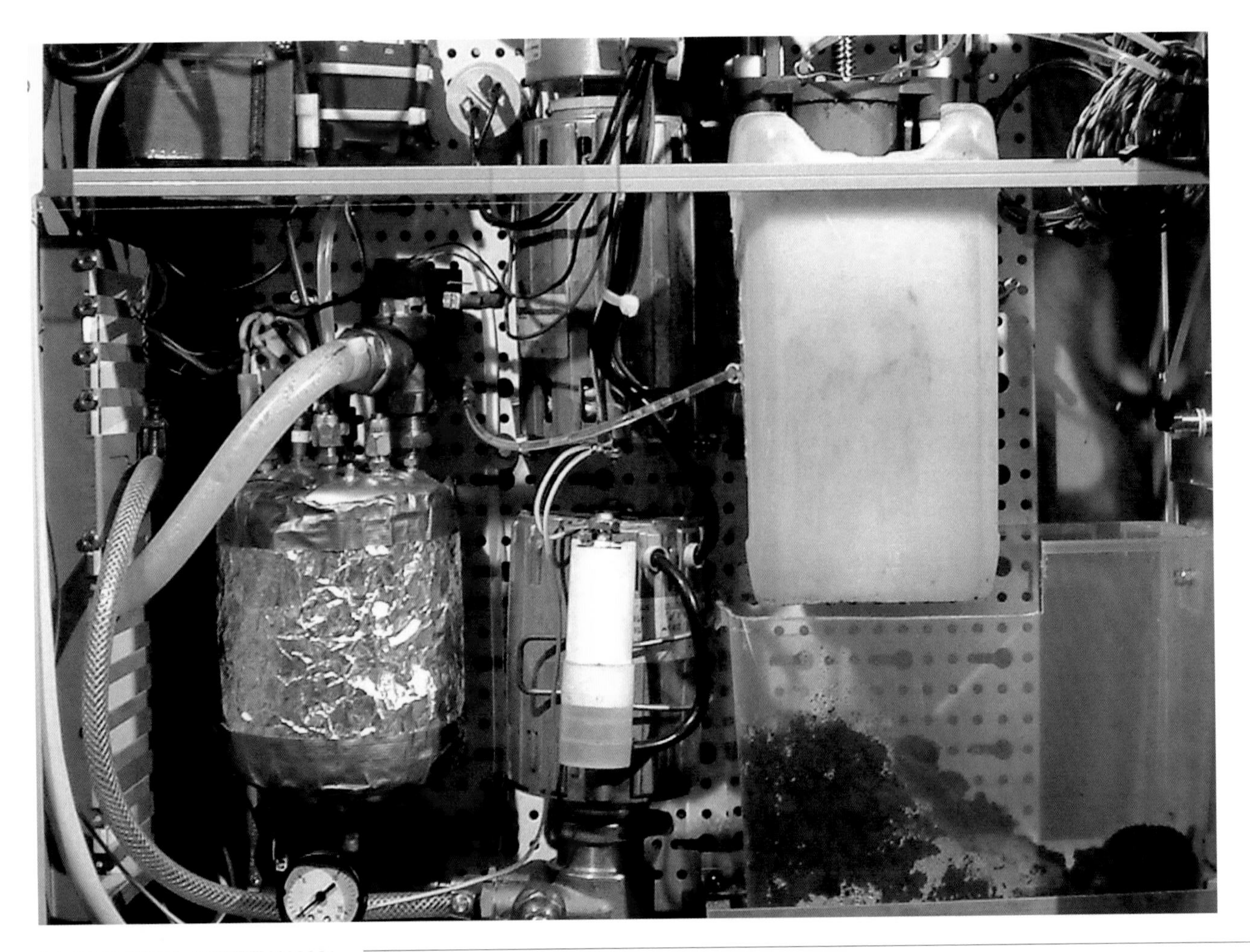

Grau pours espresso from the coffee machine he built into a computer enclosure. A cold water hose screws into the back; in front, there's an 8-digit display to choose temperature, strength, and size of dose.

Java Development Environment

Espresso machine disguised as a PC

MAKER	**René Grau**	COST	€3,000 ($3,600)
OCCUPATION	Technical editor	TIME	80 hours
LOCATION	Munich, Germany	MORE	rene-center.de

There may be a few things René Grau, 38, neglected to disclose on his employment application. For one, before his current job as a magazine tech editor in Munich, Grau had built some of the most advanced coffee machines in the world. For another, he drinks as many as three cups of joe before starting the day and suffers from an addiction to premium Italian roast.

That's why when you enter Grau's lab full of superfast tower computers at *PC Professionell* magazine, one machine delivers a big jolt. Grau created a networked, self-diagnosing, self-cleaning, digitally controlled espresso machine, built into an unassuming beige PC case. "You can't really tell the difference between this tower and the others in the office," says Grau. "Except this one smells better ... and it makes the best espresso I've had."

With a few buttons, Grau can choose an overwhelming number of variations, including regular brewed coffee, espresso, cappuccino, and macchiato. The eight-segment LCD lets you dial up large and small sizes, weak or strong strength according to the amount of coffee used, plus specific pre-brew steps to make the coffee even stronger. (Cappuccino drinkers, bring your own steamed milk.)

Grau, who went to school in Italy to fix coffee machines, rescued the old PC case from the office dumpster, then began an eBay spree that included a coffee-maker water pump, brewing unit, and adjustable, stainless-steel grinder. The coffee reservoir on top came from a plastic milk jug. The fancy LCD and micro-controller came from a manufacturer of high-end coffee machines for hotels and restaurants. Once you pour beans in, the stainless-steel grinder pulverizes them, a motor tamps the grounds, and a kettle brews them in water up to 205 degrees. When the beverage is ready, a solenoid ejects the grounds to the trash to make way for the next cup. If you can't be bothered to program the machine in person, anyone on the office network can order coffee through an online interface.

Grau's coworkers are generally happy the machine is at the magazine. His teenage daughter, also a fan of Italian-style coffee, has visited it and approves. As long as you continue to deliver the 3,000 watts it needs to run, it can pump out over 2,000 cups daily. That's more than even Grau needs in an average day.

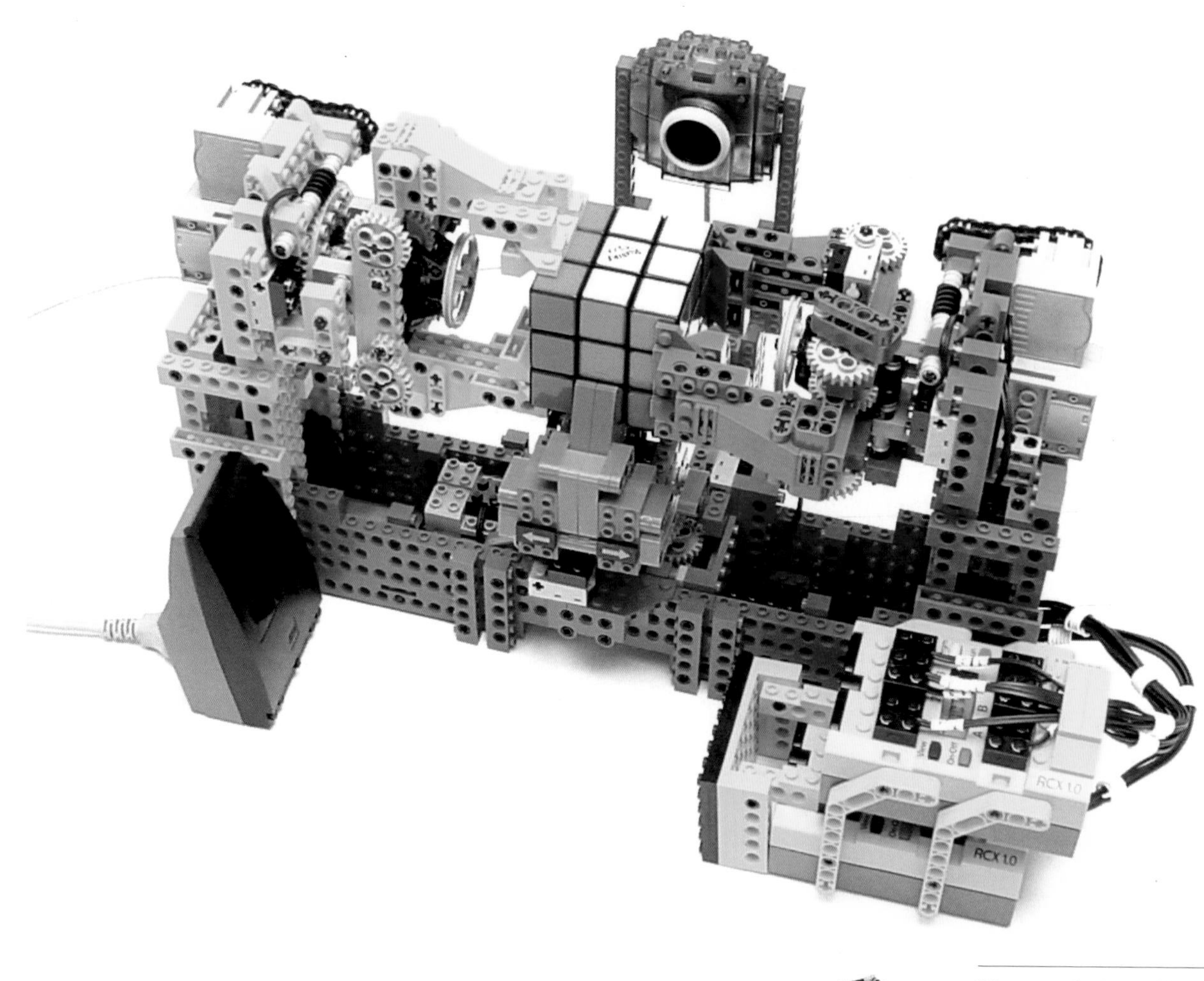

Where was he in the 80s when we needed him? Brown's CubeSolver finishes a Rubik's Cube in less than 10 minutes. The all-Lego device receives its instructions from a PC via an infrared connection. His juggling machine (left) keeps three foam balls aloft when paired with its other half.

Block Head

Machine that solves Rubik's Cube, made entirely from Lego bricks

MAKER	**J.P. Brown**	COST	$400
OCCUPATION	Museum conservator	TIME	6 months
LOCATION	Chicago, Illinois	MORE	jpbrown.i8.com

Like an Olympian shaking off drug charges, J.P. Brown wants people to know that he's never been a grinder. A Lego builder who holds a day job as a conservator at the Field Museum in Chicago, Brown explains that grinders modify the plastic Lego bricks. "There are electrical engineers who solder sensor arrays onto pieces and make very complex stuff, but to me that seems like the tail wagging the dog," says Brown, a 41-year-old naturalized American with a crisp British accent. "The coolness factor comes from having to live by the constraints of the blocks."

One of the top Lego builders in the world, Brown's most famous creation is 2001's CubeSolver, the first Lego robot to finish the Rubik's Cube puzzle. Once you place a messed-up cube in the bot's rubber grippers, its optical sensors signal the color values to the central processor, or RCX, in Lego-speak. Brown programmed algorithms for making the moves in a hacked Lego programming language called NQC or Not Quite C. Grippers twist the cube through each step and solve the puzzle in about ten minutes.

Brown has also created a robot dog that can retrieve a ball, a pair of juggling hands that can keep three foam balls in the air, and a missile launcher that tracks objects optically and fires plastic ballistics at coworkers. "It's like object-oriented software; you break down a complex job into simpler individual processes and go from there," he says.

Brown didn't work with Legos until he was 34. He and his wife were at a party with friends, and he spent much of the night discussing the Lego Mindstorms robotics kits with a kid at the party. It was 1998, and Mindstorms fired up his imagination. "Just seeing the potential was really exciting," he says. He immediately bought a set, but was disappointed with the accompanying software. Years later, he discovered the NQC language, and he pulled the toy out of the closet. "I spoke C, so I felt like I was in my natural environment," he says. "That made all the difference."

Now, Brown, who moved the U.S. from England when his wife got a post as a history professor at the University of Chicago, keeps more than half a million pieces organized in bins around their apartment. He's teaching a few insider tips to his son, Rush, 9, who's already built a solar-powered wheel that fills a room with refraction patterns. His job as a conservator — a little bit of materials science mixed with archaeology, he says — involves fixing up items from archeological digs. Plus, he has a few contract gigs with Lego on the side. After making the cube solver, Lego commissioned him to make copies for trade shows around the world. The builder asked to be paid in bricks, sending an email request for over 4,000 of the toughest-to-obtain parts.

Brown's still happy working in his chosen artistic medium, but it isn't always easy. There are limits to the material, including the strength of the plastics and the overall design of the pieces. Another problem is speed. He recently gutted a machine, which, when given a new sheet of paper, folded and launched a paper airplane. It actually worked but was painfully slow (2.5 minutes per plane), and turned out two duds for every successful flight. Lego motors have such low torque that they must be geared way down to do any real work. "It was just amazingly boring to watch," he says. "I'm not completely done with the idea, but I've put it on the shelf for a while."

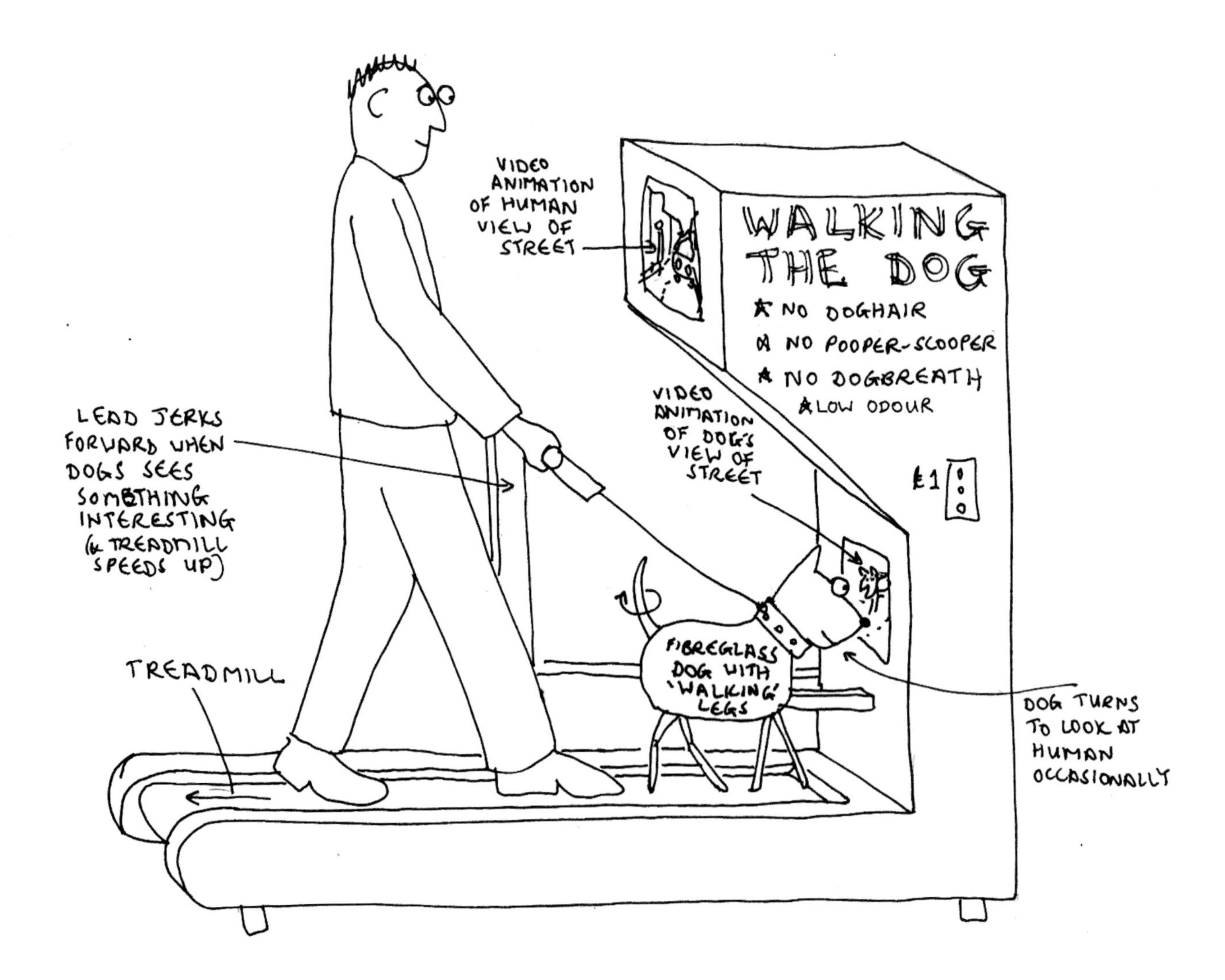

Arcade visitors in Southwold Pier pay to be pulled by a fiberglass mutt on a video-mechanical contraption 6 feet tall by 7 feet long. As a teenager, designer Hunkin worked for an amusement machine manufacturer.

Pooch Power

Arcade dog-walking machine

MAKER	**Tim Hunkin**	COST	£6,000 (about $11,000)
OCCUPATION	Cartoonist, engineer, and arcade game designer	TIME	3 months, spread out over a year
LOCATION	Suffolk, England	MORE	timhunkin.com

On the whole, video arcade management isn't usually a calling for the high-minded and scientifically curious. Just try to get your coins back from the guy behind the plate glass by claiming Centipede ate 'em.

But here's Tim Hunkin, 51, making a happy trek to work, from house to barn workshop. He passes a garden with a 15-foot hand-cast statue of Michael Faraday and a giant sundial. The barn itself features a dusty thatched roof ("Terrible for welding," he notes) and a neighborhood ginger cat, Percy, who usually noses around while he works.

Hunkin makes his living from a small arcade on England's Southwold Pier, packed with the 18 bizarre games that he created with friends. Visitors to the pier drop in between 20 pence and 2 pounds sterling per try, and revenue is cut 50/50 with the pier's owner. It's the best balance he's found between creativity and commerce. In his past, Hunkin has been a TV host for the cult series *The Secret Life of Machines*, a newspaper cartoonist, and a professional sculptor. The latter didn't suit him. "I found the hushed, respectful atmosphere of art galleries arid compared to the boisterous, cheeky atmosphere of fairs," he said in a recent lecture.

Nothing could be more cheeky than Rent-A-Dog, in which visitors are pulled along a treadmill for a dog's-eye view of the world. The human watches an animated view of the Southwold Pier on a large CRT above, while the dog watches a smaller LCD below. If the dog notices a bag of potato chips in his screen, the treadmill jerks you to a stop and you have to yank hard on the leash to get it going again.

The project started when the arcade's landlord gave Hunkin the broken exercise machine from the pier's gym. Hunkin fixed up the motors and changed the gear ratio to slow down the machine considerably ("I didn't want a lawsuit for going full speed"). He machined a steel coupling so that it would fit the existing shafts.

Hunkin has an electrical engineering degree from Cambridge, but he says his studies were mainly theoretical, and he had to learn actual electronics from Forrest Mims books just like everyone else. Now, his collaborator, Graham Norgate, who lives a few miles down the road, often completes the electronics, working off Hunkin's hand-drawn sketches of the circuit designs. Instead of using typical microcontrollers from Atmel or Microchip, Hunkin chooses factory automation chips from Mitsubishi called Programmable Logic Controllers, small boxes that accept direct connections to relays and sensors. At $200 to $700, they're much more expensive than the $1 alternative, but Hunkin claims they're more dependable.

To create the animated landscape in the video, Hunkin shot digital pictures of the familiar buildings of the Southwold Pier. Using a large-format inkjet, he printed them out around 15 inches high on card stock, and folded them into 3D buildings on a large table. He shot a video tour through the paper dioramas. The effect worked perfectly, and comically. At one point, barn cat Percy makes a serendipitous live cameo across the paper city.

As soon as Rent-A-Dog was done, Hunkin invited his partner Meg and a few neighborhood kids for a road test. The videos, he thought, ran way too long, but everyone seemed entranced. "Flat photos on a screen can look surprisingly real," he realized. "It creates a world that you just about accept."

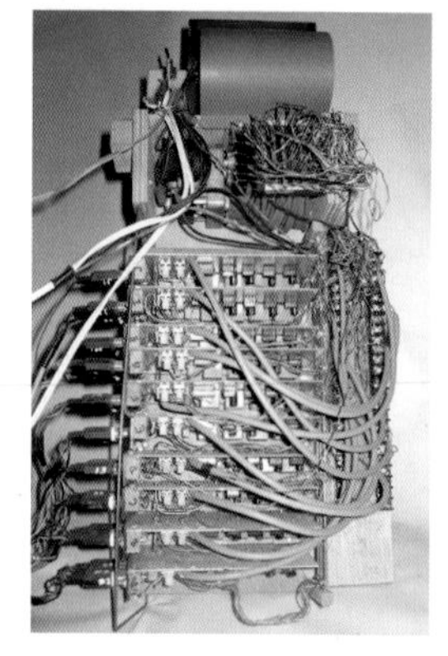

Cotton feeds music scrolls made from inkjet transparencies (top) into the instrument's optical sensors. Aluminum tubes from a boating supply store form the resonators (far left). Each of the 37 mallets has its own analog control circuit (near left).

Lionel Hampton on Autopilot

Player marimba with optical reader

MAKER	Larry Cotton	COST	Several hundred dollars
OCCUPATION	Engineer, math teacher, piano player	TIME	About 1 year
LOCATION	New Bern, North Carolina		

Talk about stubborn: Larry Cotton, a semi-retired engineer, didn't want PC hardware anywhere near his elegant marimba, even though it would have made things a lot easier. Most makers connect instruments to a computer via MIDI or some other common interface. Cotton had to rig up a low-tech optical reader inside the marimba to play custom piano rolls. "I was just kind of doing my own thing, committed to use nothing that existed already," says Cotton, with the kind of elegant Carolina accent you don't much find outside Hollywood movies about baseball.

He started the project by experimenting with photodiodes and 300-pound fishing line. He directed the beam of a spotlight through the thick monofilament, creating a cheap fiber optic system that delivered tiny spots of light on transparencies used as musical rolls. He lined up 40 such spots across 4 inches of makeshift roll, each one accounting for one of the 37 handmade oak bars around the circular instrument (with three left over for future use). An inkjet printer furnished black dots on the transparency, and when the dots blocked light, a voltage change from the photodiodes registered to Schmitt triggers and 555 timers. Each hand-soldered circuit drove relays and higher-current solenoids (former cassette player mechanisms bought surplus).

There was one problem: no music rolls existed that would play in Cotton's optical machine (old player pianos use pneumatic systems and modern pianos use CD-ROMs). Cotton makes the media himself by playing songs on his piano (he's played every Wednesday night at a local country club since the mid-80s). By capturing the audio on his workshop PC and running it through Cakewalk 6.0 software, he's able to reformat the MIDI computer dots to play in his machine.

The oddly shaped marimba pounds out a mean version of "Flight of the Bumble Bee," along with the old Carmen Miranda tune "Tico-Tico" and other songs. Christmas fare is a big hit. But like nearly all player instruments, you can hear the difference. Asked if there are times you can close your eyes and almost hear a human player, Cotton laughs. "Oh, yeah," he says. "A pretty bad one."

"Works flawlessly," says the creator, who uses a handmade vinyl pouch and plastic tubing for his $5 bagpipes. After drilling fingerholes into the chanter pipe (center), he fills it with Bondo adhesive and graduates it to a narrow circumference inside. All the fittings are secured with waxed dental floss.

PVC Piper

Bagpipes from plumbing supplies

MAKER	**Dennis Havlena**	COST	$30
OCCUPATION	State park musician	TIME	1 week
LOCATION	Cheboygan, Michigan	MORE	ehhs.cmich.edu/~dhavlena/bagpipe.htm

Want a full set of Scottish bagpipes for only $20? How about a hurdy-gurdy for $30? Ever wanted to try the stand-up bass? You can for just $12 (sorry, for electric bass, please add $3). They're all free blueprints offered by Dennis Havlena, a designer of dirt-cheap, high-quality instruments on the web.

"A bass is too expensive for most people, but with a little effort you can get first-class sound," says Havlena, 58, who works as a musician by day for tourists in the Michigan State Parks system. He says his instruments made from common hardware store parts can help people learn the real thing. "I made my bass with a cookie tin, a slab of pine, and some weed-whacker strings. Anyone can build something that plays good, sounds good ... but not necessarily looks that good."

His bagpipes, for example, are legend on the web and show a masterful command of Naugahyde, Bondo, and PVC tubing. Havlena joins two pieces of the Naugahyde vinyl from a fabric store with silicone glue to make the bag. Cold-water PVC pipes are sleeved with various thicknesses of aquarium tubing and terminated with reeds from a music store. All fittings are made with windings of waxed dental floss, and a clever check valve is made with a rolling pin, wax paper, and gob of silicone glue.

Havlena's first budget instrument was an ersatz hammered dulcimer he made more than 25 years ago when he didn't have enough money for the real thing. When he moved on to bagpipes, he was already playing a traditional set of Highland pipes. Because self-made instruments are fun, he found that the homemade backup set of pipes helped him practice more. "One day I went out in the field behind the house to keep the neighbors from shooting me," he recalls. "I'm playing on this thing, and I got enough breath to play a whole tune. I remember thinking, 'Good lord, I'm a bagpiper!'"

All that practice paid off when he got job at a restored British fort on Mackinac Island, where he runs a small department of musicians who bring traditional instruments, antique guns, and furs around to elementary schools during the winter to teach the history of the area. During the summer, they perform music and firearms demonstrations at the fort.

All his instructions are clearly written on the website. But when his internet provider's one-megabyte limit started to cramp his style, Havlena and Emily, his 10-year-old daughter (he's a late-in-life dad), thought up an ingenious way to keep the site live and make a little money. They removed large photos of the instruments and the MP3 sound files from the site and burned them onto a disc. Many people apparently are willing to pay an extra $9 to have the multimedia mailed to them (Emily takes a 25% cut to mail them out). Since 1997, Havlena has pulled in $2,000 to $3,000 per year that way.

Apparently, word is spreading. "At work, I play the bagpipes, and then we do a little dance, and then I play the fiddle," recalls Havlena. "This guy comes up to me, and we start talking, and the guy says, 'Yeah, you know there's this nut on the internet that tells you how to build a bagpipe out of a plumbing pipe!'"

Surf and Turf

Barbeque-powered pool heater

MAKER	**Todd Harrison and Veronica Harrison**	COST	$250
OCCUPATION	Database administrator and college instructor, student	TIME	40 hours
LOCATION	Mesa, Arizona	MORE	members.cox.net/toddharrison/Grill_Heater/index.html

Forget your preconceptions of sunny southern Arizona. It's way too cold to jump into your average suburban pool after Labor Day. "My relatives would come for Thanksgiving and Christmas and expect to swim," grouses Todd Harrison, 37, a programmer for a property management company. Irritated that his pool was only usable during the three months of summer, Harrison began thinking of quick and dirty ways to fire it up.

He went to Home Depot and bought 180 feet of pipe, elbow fittings, and a garden hose. He enlisted 13-year-old daughter Veronica, an experienced tinkerer of robot kits and model rockets. Together they would create a heater element that could slip in and out of the family's Char-Broil grille, using the pressure of the pool hose to circulate water. If Harrison's North Dakota State University physics skills were up to snuff, the system would realize a 7°F increase in 8,000 gallons of water for every 5-gallon tank of propane. "I had some doubts," says Veronica, "But my dad has things mostly figured out before he starts something."

Veronica cut and cleaned the pipe with an emery cloth, and her dad sweated the pipes with 80 fittings. Growing up in North Dakota, he learned to solder fittings from his father, who owned a heating and air conditioning business. Using an air compressor and pneumatic cutter, he sliced a window in the back of the grill to get more air circulation.

It worked. One chilly spring weekend, the Harrisons brought their pool up to 92°F. It's flexible enough to set up for a weekend party. "We can take the coils out of the grill and be back to cooking hamburgers in less than three minutes."

A grill loaded with 180 feet of pipe — carefully prepped by 13-year-old Veronica Harrison and soldered by her dad — circulates hot water back to the family's heated swimming hole.

Run Silent, Run Deep

Vegetable oil-cooled home PC

MAKER	**Markus Leonhardt**	COST	€25 in vegetable oil (about $30)
OCCUPATION	Student	TIME	4 hours
LOCATION	Wilferdingen, Germany	MORE	oilcomputer.com

Markus Leonhardt needed to make his PC quieter. While he studies engineering at Fridericiana University, Leonhardt, now 21, lives at home, in a small bedroom with the cooling fans buzzing next to his head all night. Over beers one night (oh, those German drinking laws), the young man and a friend cooked up a plan to effectively dampen the fan noise in his PC.

Leonhardt bought six and a half gallons of vegetable oil at the supermarket the following day. He requisitioned an old family fish tank from the basement (he once kept groupers but they died), and he removed the hard drive. He tossed the computer into the aquarium, poured in all the vegetable oil, and switched it back on. Lo and behold, the machine fired up without a hitch and has run for a year. The only problem he's experienced in the last 14 months? "There's a little bit of a smell when you open the cover," he admits.

What Leonhardt and his friend knew, of course, is that most oils do not conduct electricity. "My main goal was to make the computer silent," he says in excellent English, but with a voice that's improbably low. "In addition, I just needed something to do with my time."

The simple experiment prompted hundreds of threads in forums on Leonhardt's website. The student, who eventually wants to practice electrical engineering for power plants, is overjoyed at all the attention and is planning his next "oil computer." He wants to immerse his 2GHz server, perhaps overclock the processor to get better performance, and provide an external radiator to counter increased heat. He's a little superstitious about keeping the medium the same, however. "Everything works with this type of vegetable oil. I don't think I'll change."

Slippery slots: It turns out the 21-year-old's PC ran a lot better after he sank it in 6.5 gallons of cooking oil.

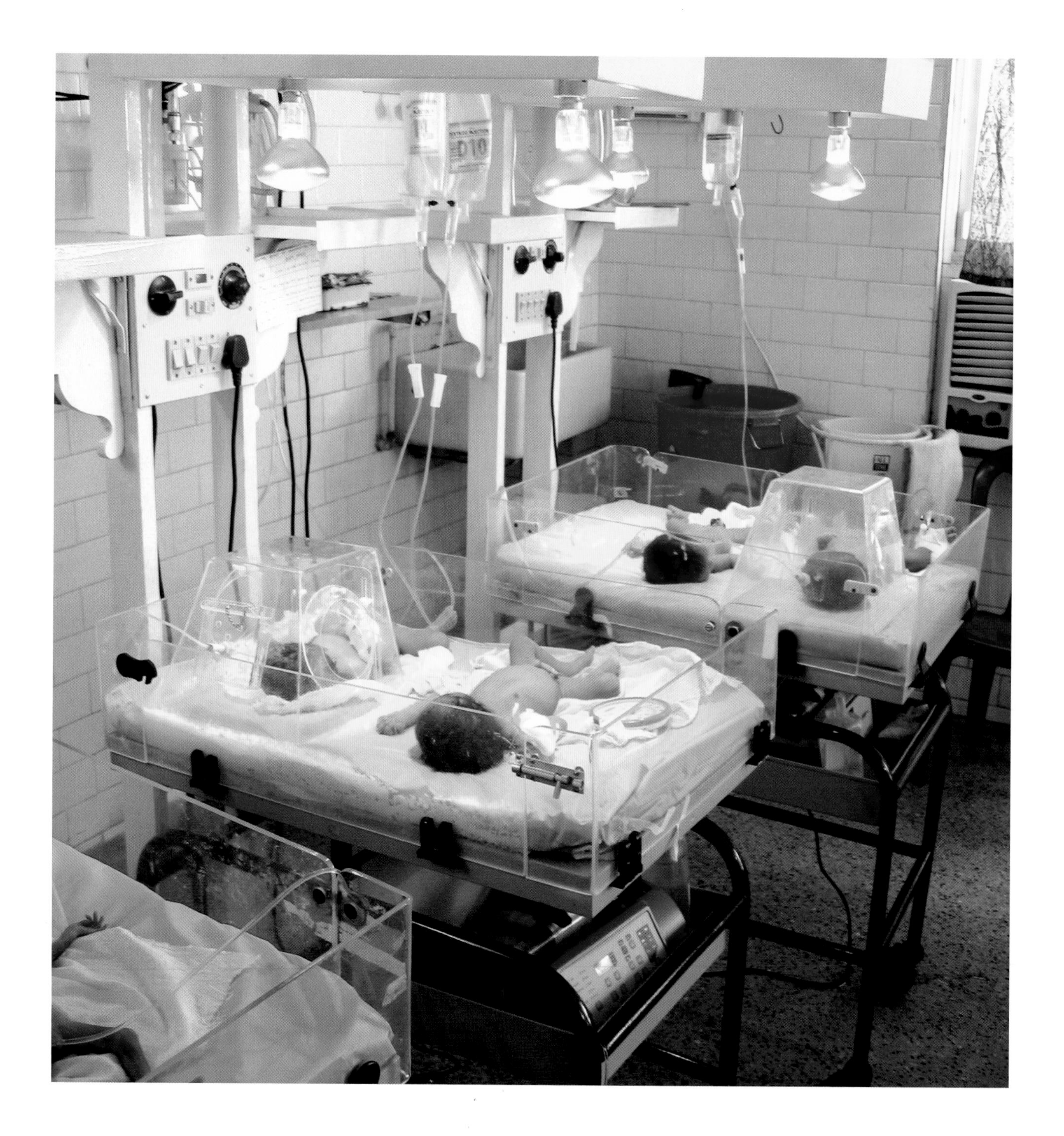
D10

Cheap Heats

Improvised baby warmer

MAKER	**Sathya Jeganathan**	COST	Rs4,500 ($100)
OCCUPATION	Physician	TIME	3 days
LOCATION	Chengalpattu, India		

After crunching a few numbers, Sathya Jeganathan, a 42-year-old neonatologist, noticed an unusually high infant mortality rate at her workplace, a government hospital in India. Of the thousands of children born annually at the underfunded facility, an alarming 39 out of 1,000 did not survive. One major problem: the hospital lacked infant warmers, those raised plexiglass beds common in most birthing centers that help newborns regulate temperature.

Jeganathan, who earned her medical degree in Italy and her master's in public health at Johns Hopkins, immediately wrote a grant to have warmers brought in. A few devices arrived at the hospital, but they had sophisticated temperature alarms and were more trouble than they were worth. The new systems — which typically cost $4,000 and up — required a human operator to turn on the heat manually when the baby's temperature dipped too low. "There's one nurse to care for 20 to 30 babies, and the alarms were going off all the time," she says. "When they broke, no one was willing to come from town to fix them."

So the doctor took the neonatal staff nurses out for a day at a nearby beach to discuss the problem. Babies have a tough time keeping a constant temperature because of their high surface-to-volume ratio. And it's even worse for premature and low-birth-weight children. Commercial incubators use radiant coils — like a stovetop — to generate heat. What if they could use the heat of light bulbs instead?

Carrying rough drawings, she approached city electricians to help her cobble together a prototype. At the time, wood was cheap in the surrounding villages because a recent storm had blown down trees. The first unit consisted of a table with a 100-watt bulb above the head of a child, and the electricians watched as the device helped a 2.6-pound preemie grow to a healthy weight. "We all developed good bonding after that," says Jeganathan.

The informal group built several more structures and fitted them with multiple light bulbs, purchased at local markets. The electricians installed dimmer switches that can be turned down to allow a baby to sleep at night. They put in a dial that switches between the two warming lights and an examination light.

Now fifteen improvised infant warmers operate around the clock, and Jeganathan says the mortality rate has been nearly cut nearly in half to 22 in 1,000. At most, the device costs $100, with a fancy fluorescent version for jaundice for about $150. As a side benefit, the lights help the busy nurses make a quick visual check on children from across the room.

Jeganathan's telephone manner — soft-spoken and at times chatty — belies her uncanny ability to carry out big changes. She's currently working with her electrical team on a television system in the waiting room to teach child-care tips to new moms, and is planning a solar-powered sterilizer for medical instruments.

She says she draws inspiration from her parents, social workers who still put in 12-hour days. Her father, 90, once worked with Gandhi; her mother, 80, advocates for the poor in nearby Nagapattinam, India. For her part, Jeganathan evangelizes the warmer's design outside of the hospital. The team has already shipped six warmers to nearby newborn units, and presented her designs at an international medical conference in Madrid. Finally, a positive spin on global warming.

Zvika Netter (top) at the wheel. A DLP video projector over his right shoulder displays the game, while full-sized stereo speakers on the hood play the soundtrack through the windshield. Netter and Tal (left) started the project in their Tel Aviv workshop.

Road to Nowhere

Racing simulator in an old Renault

MAKER	**Zvika Netter and Yuval Tal**	COST	NIS1,388 ($300)
OCCUPATION	Software entrepreneur, software development manager	TIME	3 months
LOCATION	Givaataim and Rehovot near Tel Aviv, Israel	MORE	realvirtualcar.blogspot.com

It was a marriage made in heaven. At a three-day camp for adult geeks in Israel, called KinnerNet, Zvika Netter, 33, and Yuval Tal, 31, realized that they had intermeshing skill sets. Tal, a programmer at Amdocs, showed his new friend an old-school game console he created by hand. Netter, founder of a financial and insurance software company, took Tal joyriding in a three-wheel, water-cooled, *Mad Max*-style dune buggy he'd welded together from street detritus. They exchanged phone numbers and decided to build a monster project that would match machine code with metalwork.

In a storage area of Netter's dad's wood workshop, Tal and Netter designed their racing simulator to be as real as possible. They combed local junkyards and found a wrecked Renault for $200, delivery included. A truck dropped it off in front of the workshop the next day, and Mr. Netter, Senior tucked it inside with his forklift. They excised the engine and sliced out much of the sheet metal inside the car using a circular saw.

Once they had access to the steering column and instrument panel, they hooked up parts that would interface with a PC running NASCAR Racing 2003 software. The racing sim was used because it is one of the few that provides telemetry data to programmers such as speed, rpm, and distance traveled. To create the steering wheel, they used a Logitech game wheel. Netter machined a steel coupling for the wheel and welded it to the car's steering column. They created parts that would protect the Logitech's delicate parts from over-rotation, but had to abandon the force feedback because it would not move the heavy automotive parts. Tal began writing the code for an Atmel microcontroller for the game's interaction with the wheel and dials.

At first, the partners scooped out all the original actuators in the Renault's speedometer and tachometer, replacing them with servos. But the little motors made the needles jump between steps, ruining the dashboard's realism. Netter was not happy, and Tal developed a controller that would power the car's original electronics, which work by magnetic coils. "It's one of the coolest things in the car for me," says Netter. "If you drive a few kilometers, you can actually see it on the mechanical odometer."

A white sheet on the windshield serves as a makeshift screen, with a DLP video projector casting the image from behind the driver. Two large, hood-mounted speakers blast the game's soundtrack through the sheet and provide great thump.

All reports indicate that the real-fake car is very convincing. Netter's 3-year-old son, Omer, likes the illicit feeling of driving an adult's car. And Tal's wife, Pnina, drove it with a sense of realism that left him a little baffled. "She was driving so slow, and I said, 'Come on, it's a game, you should drive faster, do something!'" His wife replied: "No, I'm a careful driver, I like to go slowly."

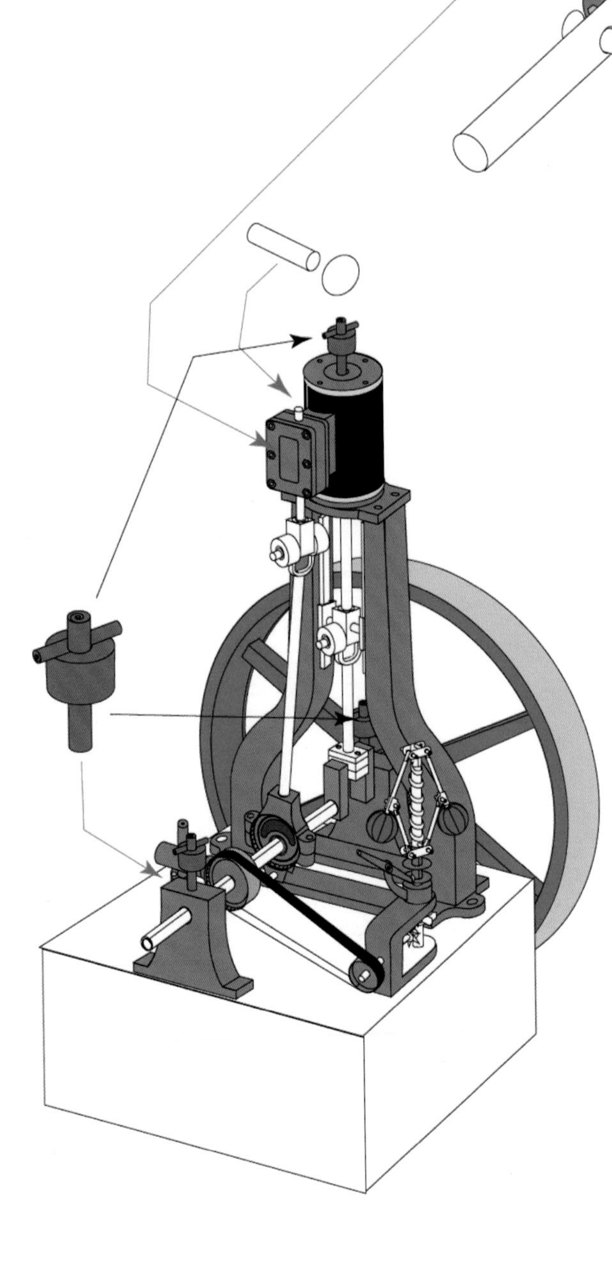

Steam engine blueprints describe assembly of pieces cut from card stock and laser paper. Below, a flywheel governor controls the airflow into the model, providing feedback to the engine. Complete plans are free on Bertsky's website.

Industrial-Age Origami

Fully operational paper steam engine

MAKER	**Ed Bertschy**	COST	$7
OCCUPATION	Graphic artist	TIME	1 year to design, 14 hours to build
LOCATION	Tucson, Arizona	MORE	home.earthlink.net/~eellbee/steam2.html

"The first time it turned over, I got a thrill out of it," says Ed Bertschy, 51, a graphic artist and maker of paper models. Bertschy had piped a little compressed air into the piston of his 1870 engine replica, and the little paper crankshaft worked up and down turning the flywheel and power wheel. The paper governor started spinning fast, pressing two weighted balls by centrifugal force to regulate the speed of the engine. "That was the aha moment," he says. "Then I was absolutely obsessed with it. I had to finish and get all the pieces to work."

Most of the parts on Bertschy's paper model work just like the real thing. On an original Riches and Watts nominal 2½-horsepower vertical A-frame double acting simple slide-valve steam engine (try saying that a few times), the flyball governors represent some of the world's first feedback devices, keeping engines from exploding from excessive speed. With the paper version, they just look really cool. All done with a few sheets of paper, a bottle of Elmer's, and a #11 X-Acto.

Bertschy started getting into paper modeling in high school when he built an intricate F-4 Phantom that successfully glided the length of his parents' hallway. Since then, he's done a cello, dynamite plunger, and 1960s ramjet missile drone. Bertschy sees paper as one of best prototyping materials, and points out that the glue dries quickly in the arid Tucson air. He's very principled about sticking entirely to paper. "Some people use balsa wood or Q-tips for shafts, but I roll them from paper or cardstock," says Bertschy, who sports a small ponytail and works by day on promotional materials and diagrams for an avionics manufacturer.

Paper has always been Bertschy's medium, professionally and as a hobby. Once he got out of the Army after high school, he started a freelance graphics career that took him into several uncharted print media. At the time, he stayed home with his two daughters while his wife worked full time. One job required him to create scratch-and-sniff lottery tickets. Another put him in charge of vegetable-dye tattoos for kids. But Bertschy settled down with the aeronautics company nine years ago, sticking to models and a few other hobbies. In 2003, he designed the special effects and props (all in paper) for a $100,000-budget indie film about a guy who fixes up an old rocket and flies to the moon. With his brother, Bertschy's currently building a full-size, fully operational, steel-and-wood replica of a 1912 Martin-Curtiss pusher bomber airplane.

The steam engine project started when Bertschy saw that a Norfolk, England, man had meticulously disassembled and diagrammed a Riches and Watts machine, once used for pumping irrigation water. After creating a satisfying paper model for himself, Bertschy formatted all the parts in full color in Illustrator, and placed them on the web for free. His assembly instructions are clear and witty; six people have made the model and added their own parts, like electric generator replicas.

When he finally showed the steam engine to the Norfolk man, the man groused that the flywheel arms were the wrong shape. OK, so paper can't do "curvilinear surfaces," says Bertschy. It's still a worthy imitation. And, sure, there may be a few other downsides to paper. The material doesn't seal well against air leaks. Now some pieces have to be turned by hand to see them move. And then there are paper cuts. "I really do get them," he says. "There's a little sweat and blood left on everything I do."

Dunklee on his modified electric scooter. The solar panels fold on piano hinges, stowing neatly to the sides for driving. The vehicle hasn't required an electric charge in more than 900 miles.

Sun Hog

Solar-powered scooter

MAKER	**Don Dunklee**	COST	$3,300
OCCUPATION	Drugstore manager	TIME	80 hours
LOCATION	Davison, Michigan	MORE	builditsolar.com/Projects/PV/pvscooter.htm

For 27 years, Don Dunklee has kept a perfect-bound school notebook tucked into his desk at home. It has a black-and-white marbled cover, tattered edges, and sketches of every engineering brainstorm Dunklee, 45, has had since high school. "Sometimes I'll sketch something out and date it, and then add to it later," he explains. "It's mostly random thoughts, silly things."

A few years ago, Dunklee reviewed an old sketch of a three-wheeled solar car, and decided to try to make it. Instead of creating the vehicle from scratch, however, he bought an electric scooter and attached solar panels.

The results aren't far from Dunklee's original vision. The scooter has traveled 900 miles in five months without needing a jolt from its plug. Every day, Dunklee drives it ten miles round trip to his job as a manager of the local Rite Aid. During the commute, the four 30-watt solar panels fold neatly to the side; when he arrives at the drugstore parking lot, they unfurl and soak up the rays like a big prairie sunflower.

Dunklee admits the bike makes him a little conspicuous in Davison (official slogan: City of Flags; population: 5,536). "When you first ride into town on a bike like that, everybody is wondering what the heck it is," says Dunklee. And yet, he's happy to explain the financial benefits: zero fuel costs, a $75 insurance charge, and a low three-year registration fee. One senior citizen insisted on a test drive once she heard his pitch.

The scooter Dunklee bought for the project was an EVT 4000e. In the trunk, he mounted a Xantrex C40 solar charge controller, which evens out the voltage from the panels (the bike's expensive gel batteries would be ruined by a voltage spike). He then soldered the panel feeds together in series to the controller, and spliced the controller output to fuses and the 48-volt battery feeds so that the bike can also easily be charged from a standard outlet. To create brackets for holding the panels in place, Dunklee used ¾-inch iron pipe, stock aluminum, and wide piano hinges. He hired a welder to fasten it all securely to the frame (total charge for welding services: four tubes of VapoRub from the pharmacy).

The solar panels offer only a tiny charge at any one time, but it's enough for Dunklee's commute. He drives the vehicle under 25 mph to minimize the load, and is careful to direct the panels. "I go out in the parking lot every few hours and give it a kick in the direction of the sun," Dunklee says. "It's like cheap solar tracking."

Despite the obvious financial benefits of riding such a vehicle, Dunklee says his kids, ages 18, 20, and 22, "don't want to be seen with it." (The family also has a Jeep and a Saturn for group trips.) Dunklee and his wife still live in the same 1,500-square-foot farmhouse they built themselves a year before their first kid was born. Thanks to the solar panels on the roof, they have lived without power from the utility company ever since. (When asked to explain where his interest in solar energy and self-sufficiency originated, he struggles for a moment. His father was an autoworker in Detroit, and his mother a bus driver. Then it hits him — a sixth-grade teacher once asked the class to create plans for a self-sustaining island.)

What other gems might be lurking in the dusty notebooks of Donald Dunklee? He will reveal only a few. One page details improvements to the solar scooter that would sense air currents and fold down the panels automatically in the event of high winds. The City of Flags has never seen anything like it.

The Speed of Starch

Semi-automatic, pneumatic potato cannon

MAKER	**Andy Gustafson**	COST	$85
OCCUPATION	Student	TIME	6 days
LOCATION	Chico, California	MORE	xinventions.com

Andy Gustafson, 21, has made spud guns into a successful academic career. In high school, he entered his innovative homemade air cannons and won numerous science fair awards, including a $32,000 scholarship. Is there a spud-gun Ph.D. somewhere in his future?

The California State University Chico senior starts his designs with the time-honored configuration of two PVC pipes. One acts as the barrel where the object will travel when fired. The other holds pressurized air from an integrated CO_2 canister that propels the ammo. The trigger is linked to a piston valve created from hardware-store neoprene.

To make the cannon semi-automatic, Gustafson figured out how to create a bolt-like system similar to that used in a machine gun. A sliding, notched piece of PVC seals the chamber while it's fired, then slides back, grabbing a new piece of ammo from the feeder pipe. The only problem is that after the pressurized chamber fires, it takes 6 seconds to refill. Gustafson's now finished with a fully automatic version that has four pressurized PVC chambers encircling the barrel so that one can be firing a tuber while another is being filled. The gun fires 120 rounds per minute.

Gustafson claims he has never had trouble with police, not even a run-in with an overeager rent-a-cop. Since he sells the plans for some of his more elaborate guns on his website, he asked the Bureau of Alcohol, Tobacco, Firearms and Explosives about legal issues. "The Bureau has previously examined devices known as Spud Guns, Potato Guns, or Spudzookas," they wrote back, "and determined that such devices, in and of themselves, are not firearms." At least he got a federal official to use the word spudzooka.

The deluxe cannon boasts ergonomic handles and a push-button trigger. It auto-loads fresh spuds and foosballs every six seconds, launching them 300 yards. A similar prototype fired a lemon through aluminum plate.

Six Monkeys

Hand-machined merry-go-round

MAKER	**Al Gori**	COST	$1,200
OCCUPATION	Millwright, machinist, and mechanic	TIME	2 months
LOCATION	Cozy Lake, New Jersey	MORE	homespunmerrygoround.com

On weekends, Al Gori usually has an appointment around the New York area. The 41-year-old hauls a homemade merry-go-round out of his Chevy Astro van, decorates it with animatronic monkeys, dons a pith helmet, and starts the motor. Most of the time he does this for absolutely free.

After decades of taking youngsters for spins, Gori realized he actually feeds off the minor celebrity that comes from this gentle performance art. In 2005, he made a new version using the skills he's learned as a professional millwright, expert in servicing and assembling intricate gears for factory machinery. Gori rehabbed an old paper-mill gearbox to bring the attraction to life. He tapped a store of windshield wiper motors he keeps in his garage and created the moving animals. The device runs off two car batteries so he can operate it away from a power source.

Gori must smile each time he stands in safari gear, a nod to old-time carnival impresarios. After all, his first 450-pound carousel started as an art thesis for his master's in kinetic sculpture at Rutgers over 20 years ago. It was a dark and foreboding object called *The Merry-Go-Round to Hell,* made from a 1968 Dodge Dart drive shaft, a washing machine motor, a printing press gearbox, and other scavenged parts. Over the years, he added butterflies, monkeys, snakes, and other winged animals to cover the steel parts.

On a recent weekend, Gori trekked to New York City's Tomkins Square Park and then to a church picnic. "It's been helping my social skills because I'm tinkering alone most of the time up at my shop," he says. "And this is my time to get out and see people."

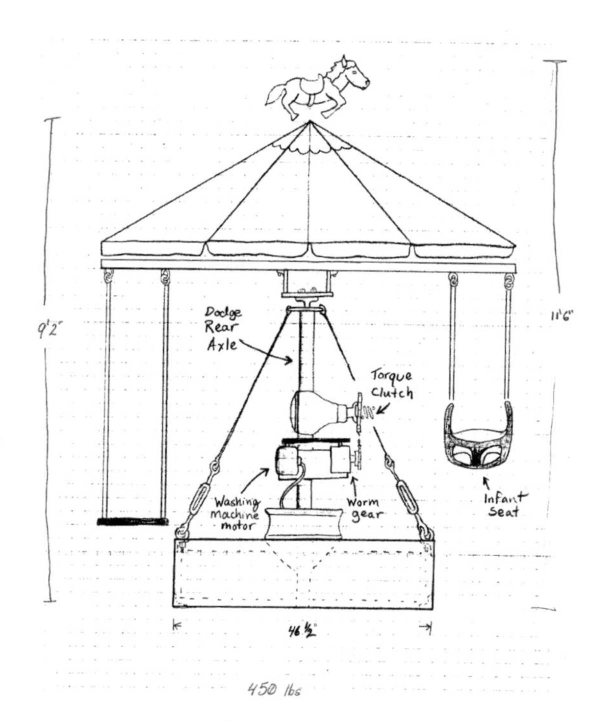

An expert in industrial gearing, Gori created a drive system for the merry-go-round using a printing press gearbox.

The nine-ball ping-pong dispenser counts down its ammunition on an LED fixed to a microcontroller board. When players press a button under the table, a motor from an old inkjet printer activates and fires a new ball.

Ace Service

Motorized ping-pong ball server

MAKER	**Dimitri Merrill**	COST	$120
OCCUPATION	College student	TIME	40 hours
LOCATION	Fort Collins, Colorado	MORE	designnews.com/article/CA513573.html

A college student has solved one of humanity's most pressing technical challenges. Who hasn't wished for an automatic ball dispenser for hairball-free beer pong?

The project was first whiteboarded in Dimitri Merrill's basement, part of a house that he rents with four other students from Colorado State University. The place is a spacious ranch perched on the edge of the Rocky Mountains, a long two-mile stretch from campus. A benevolent roommate brought a deluxe ping-pong table for the cut-off students during the school year, but troubles arose whenever the group moved to beer pong — which involves positioning your beverage in the middle of the table while your opponent tries to aim the ball at or into the cup. Players must sip or chug their drinks when the cup receives a direct hit.

"When we played, the balls would get sticky from the beer, fall on the ground, and roll in dirt and hair," recalls Merrill, 22. "I started thinking of a way to automatically have clean ones to play."

During the next semester's most challenging class, Mechatronics Engineering, Merrill's assignment was to create an electromechanical device of his choice. He knew just the project. The machine would stand in the center of a ping-pong table, supported by a PVC stand and connected to two buttons on opposing sides of the table. When a player presses a button, a new ball drops from the machine on his or her side. Up to nine balls are stored in a PVC tube above the mechanism. When the ball supply is reduced to just three, the device powers an LED to show the number of balls left. A buzzer sounds when the machine is out of ammunition.

Merrill scavenged an inkjet printer's metal cartridge and motor to handle the ping-pong balls. The apparatus pulls balls from the loading tube and dispenses them in one direction or another according to voltage from a microcontroller board sold by Xilinx. He added an infrared emitter and sensor to signal the countdown of the dwindling ball hopper, from three down to zero. Three other students — Erik Stepperud, Robert Millot, and Nathan Davidson — helped put the electronics together and constructed the housing. Officially, in the paper they submitted to class, the mechanism's purpose is to "eliminate back fatigue" and reduce interruptions to the game. Yeah, sure.

Merrill's parents are proud, though they witnessed a few disasters along the way. His father is a software engineer for the National Center for Atmospheric Research in Colorado and has always been an inveterate tinkerer. His mother, a homemaker, immigrated from Greece as a young girl. (In Greek convention, Dimitri carries her father's first name. "My brother got the name Joe," he huffs.) The first time their electrical engineer son brought the dispenser to their Colorado home and powered it up, "a capacitor exploded like a Black Cat, and the plexiglass box was full of twisted metal," he recalls. Another time he visited, the voltage to the motor was too high, sending the balls flying out of the chutes at dangerous velocities.

The malfunctioning ball server was soon fixed by supplying voltage from the school's lab power supply. Merrill realized he had been using cheap wall adapters for power, which often fail to supply the exact voltage claimed by the specs. His dispenser, which didn't have a voltage regulator, needed a near-perfect source of electricity, and then the device worked harder than a Wimbledon ball boy. His team got an A-.

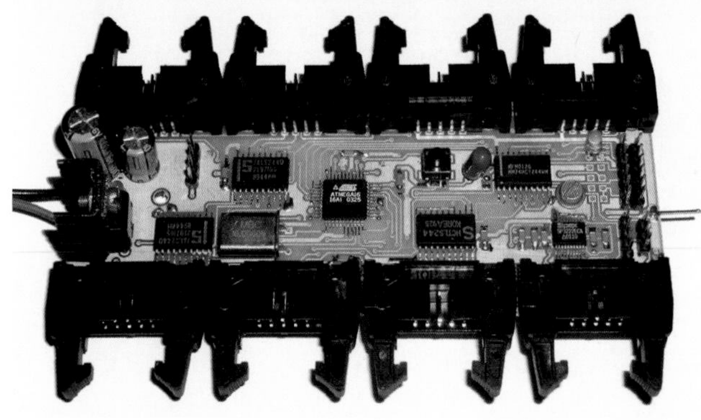

The Russian engineer drives with his homemade LED sign atop his white Volga (above). The Atmel microcontroller (far left) revolves at the center of the armature at 200 rpm. Meanwhile, a sensor (near left) checks speed by counting the number of passes of a magnet underneath.

Apparition Apparatus

Car-mounted virtual display

MAKER	**Alexander Telegin**	COST	R25,000 ($900)
OCCUPATION	Embedded systems programmer	TIME	A couple of months
LOCATION	Moscow, Russia	MORE	ledcontrol.narod.ru

Trying to get attention for your restaurant or store? Why not mount one of Alexander Telegin's amazing persistence-of-vision displays atop your car? So far, Telegin, 30, has sold eight devices which he makes out of wood and electronics with his father in a three-room Moscow flat. To drum up business, he'll get it spinning on his white Volga sedan, and periodically send messages to it from a wireless keyboard inside the car.

"It gets a lot of attention," says Telegin in an email translated into English. "Whenever I go out and set up somewhere, many passers-by stop and look." He notes, however, that live public demos are infrequent. "I drive with it only as an experiment. We'd probably have problems with the police. It's very hard to obtain permission to set up one of these devices."

Persistence-of-vision projects are fairly common among tinkerers, though not usually on such a grand scale. The devices work by turning LEDs on and off at precise moments, while they rotate at 200 rpm. Your eyes see only a text message or graphic floating in space.

Telegin, who graduated from Moscow Institute of Radio Engineering, Electronics, and Automatics, launched the project after he found a powerful motor from a computer tape drive in a Moscow junk shop. His father, a retired physics teacher, checked in from time to time with solutions to technical problems. His dad lives with Telegin and his wife (a textile industry engineer) and their two daughters, ages 1 and 3. Father and son created the arm from pine boards and then drilled holes along the outside pieces. They glued 200 red LEDs into the holes, and connected it all to an Atmel ATmega16 microcontroller mounted on the spinning arms.

To use the display, you tap out a message up to 230 characters on a battery-powered wireless keyboard, which sends the text to a receiving unit mounted on the spinning arm. You can instantly change the message at any point instantly by typing a new one and pressing Enter. The electronics on the spinning arms are powered through brushes on the motor's drive shaft and ultimately by the direct current in the car (you have to keep your car running or risk a dead battery). Some nice touches: The text is shown in a crisp Times New Roman font, and the LED-covered crosspieces feature an adjustable angle so that people can see them easily from the street.

Asked what he was thinking the moment he had it working — Was he elated? Surprised? — Telegin again voices his concern with the Moscow authorities. "I was wondering whether the police were nearby."

Fried (above, right) works the jammer on an "unsuspecting" cellphone user. The device works by employing a high-frequency oscillator circuit (left) to blanket the immediate area with junk signal.

Radio Free Boston

Cellphone jammer

MAKER	**Limor Fried**	COST	$50
OCCUPATION	Freelance engineer and artist	TIME	4 months
LOCATION	Boston, Massachusetts	MORE	ladyada.net

"It happens to me all the time," explains twentysomething Limor Fried, a recent master's degree grad from MIT. "You're stuck on a bus from Boston to New York and the person next to you talks on their cellphone for three hours ... well, it certainly feels like three hours." Fried started itching for a mass-transit solution, and soon built a pocket-sized device that could squelch phones up to 6 feet away at the push of a button.

Part of a demo for her master's thesis, the jammer uses a method called "denial of service" to overwhelm the phone's radio band with a high-power burst of junk signal. To create it, Fried bought two voltage-controlled oscillator circuits to create a radio carrier frequency. She designed a tuning circuit using op-amps and simple components. The tuner leads the VCO through all the frequencies, generated to sweep across common analog and digital cellphone radio bands from 800MHz to 1.9GHz. Store-bought amplification stages increase the size of the jamming space, and in a later version, an RP-SMA PCB edge-launch antenna allowed her to attach a variety of antennas. Unfortunately, because cellphone bands vary so widely, you have to guess which type of phone you're trying to disrupt and swap antennas appropriately. But in the end it worked with most phones. And with two AA batteries, the device can jam periodically for a few hours. Fried's grade on her thesis: A.

Though she'd like to make improvements, Fried has already moved on to her next project. She lives in South Boston with a couple of roommates and a cat named Mosfet. ("Yeah, I know, that's pretty bad," she admits.) Her electronics workshop is a door on sawhorses, on which she and a roommate make open-source synthesizer kits they sell over the internet, with sales of over 100 units at $300 apiece.

Fried grew up in both the U.S. and Israel. She credits her big sister with planting the tinkering bug. "She'd be like, 'Hey, look, 50 oil cans on the side of the road.'" Fried recalls. "She'd wash them out and make them into boxes for everyone in our family."

Despite the glee of creating a jammer, Fried doesn't exactly recommended the project to other makers. "If you want one, just buy it," she says. "They're illegal, but they're out there." (According to the FCC, it's forbidden to block licensed radio communication, though describing the device for a school project probably won't get Fried into trouble.) She worked so hard to create the device for the paper that some of the joy went out of it. "I tried it and it works, but I never got to the revenge stage."

BOUNDARY
ARGONNE WEST
RAMBLER
SLC-1
NOVA SYSTEM
Digi-Key
Perl Cookbook
FAIRCHILD

Soul of an Old Machine

Refurbished Data General computer

MAKER	**Tom Jennings**	COST	$580
OCCUPATION	Technical support	TIME	50 hours
LOCATION	Los Angeles, California	MORE	wps.com

Like a Hollywood doyenne warming up to sing her signature number, the Data General Nova 4 coughs and hiccups. The tape drive whirs and the fan noise is unbearable. This beauty really knows how to dim the lights (the computer sucks up almost 3,000 watts).

In 1978, the Nova 4 was the pride of the industry, one of the powerful minisystems chronicled in Tracy Kidder's *The Soul of a New Machine.* It has a clock speed of 4 megahertz, RAM of 32 kilobytes, and performs only if you talk to it in Fortran. Now, more than a quarter-century later, Tom Jennings, 50, has rescued it from a near-death experience. One Saturday afternoon, he received an email about a free machine lying in a Bakersfield parking lot. He rented a trailer and hauled it back to his wood-paneled workroom in the Silverlake district of L.A.

Jennings isn't a collector or a historian, but sort of a junk anthropologist. When not working on the Nova 4 or making kinetic sculptures in his workroom, he drives a refurbished 1970 American Motors Hornet with his two Peruvian hairless dogs and his partner of 10 years, Josh Stehlik. For gainful employment, he works as a tech-support guy and teacher for an interactive arts grad program at the University of California, Irvine.

Once Jennings moved the Nova 4 safely inside his cramped workroom, he knew he'd made the right choice. "When you look at it, it screams computer, like the prototypical one in all the old science fiction movies," he says. He started by removing two empty racks. A rash attempt at booting the machine before cleaning it left him with a broken hard-drive platter, which he replaced. After consulting with Data General fanatics online, he scrubbed the hard-drive heads with a new popsicle stick soaked in alcohol. Although the computer had been inside a trailer for years, it still needed a full once-over with a dry paintbrush and vacuum cleaner. Spiders were living in the printer.

Jennings knows just enough Fortran to write a few programs. He's been around computers since learning the fundamentals from his father, a self-taught technician who worked in companies around Route 128 in Massachusetts, the Nova's birthplace. Jennings is currently working on code that will enable the machine's 144 inputs and outputs to control lights in his house.

Why bring a 1,200-pound object inside your home? Jennings defends his decision by saying it's started him thinking about hardware in a new way. "The Nova 4 is simply a pleasant thing to operate. As a physical artifact, it appeals to the senses," he says, noting that this was one of the last computers in which engineers started by drawing up the actual logic gates on paper.

Jennings only works on computer projects in cool weather. He spent last summer mounting an MP3 player in his AMC Hornet. The digital controls are hidden in the dials of the car's original AM radio, the display resides in the glove compartment, and the server spins music from the trunk. AMC cars, he notes, are a great platform for hacking because all the models have interchangeable parts and they're easy to get, even though the company stopped selling in the 1980s. "These cars are utter orphans," says Jennings. "I guess I have a soft spot for orphans."

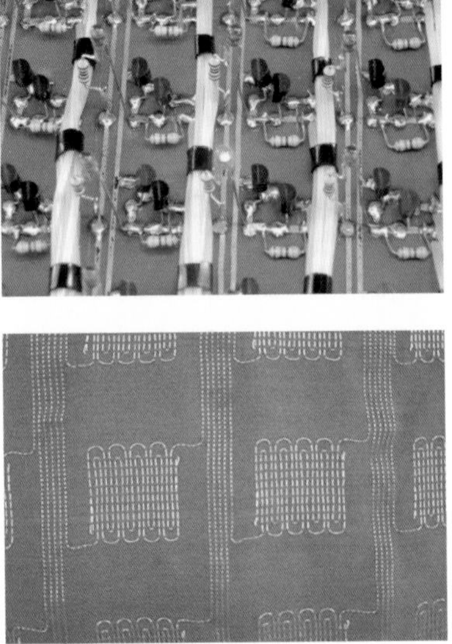

MacCary (left) at home with her loom. The conductive fibers she weaves lead to 36 amplifier circuits (top), which illuminate LEDs depending on where you touch the fabric. Her finished artwork is called *Dialectric: Connection* (above).

Shocking Duds

Hand-woven electric cloth

MAKER	**Laura MacCary and Lawrence MacCary**	COST	$1,500–$5,000
OCCUPATION	Artist, retired engineer	TIME	60–100 hours
LOCATION	Seattle, Washington	MORE	maccary.com

She wasn't exactly spinning straw into gold, but close. In 1997, Laura MacCary began loading professional audiotape instead of thread into her 150-pound wooden floor loom after she discovered miles and miles of the stuff in a nearby dumpster. MacCary, an advanced weaver and former network administrator, started to prepare it on the loom (which she keeps in her apartment) by loading the warp threads into the loom and keeping them under tension for weaving.

"The dumpster belonged to a company that sold Muzak," says MacCary, 38, recalling a lunchtime excursion she made near her former job at Aldus. "I used to check it all the time before they put a lock on it, and that day I found all this tape from a reel-to-reel machine. It was shiny on one side. I wondered if I could weave something that would act as a potentiometer." MacCary promptly brought it home and created a 5-by-5-foot audiotape blanket in a shimmery crisscross pattern, connecting the end strands to an oscillator circuit. She hoped it would produce strange sounds.

"I kept blowing transistors," she says. "But I did know one person who could figure it out." MacCary mailed off the whole business to her father, Lawrence MacCary, 80, a sculptor and longtime tinkerer. Lawrence created a new circuit around a 555 chip that would regulate the voltage and create an audio signal for the speakers. Now you can vary the tone it creates when you touch it in various places. It was a cool piece of interactivity, and the start of a collaboration in electronic cloth.

In mid-2005, MacCary and her dad developed a more mature example of the work, in a series she calls *Dialectric.* Instead of making sound, the piece illuminates 36 red LEDs according to where you place your hand. The winter before, MacCary had received her master's in fine art, and she'd been showing the piece and several others in various museums and galleries. The new piece consists of wool-polyester blended thread, which MacCary wove together with tinned copper wired in a zigzag pattern. She and her father then collaborated to solder together 36 Darlington amplifiers to correspond to each pair of wires in the piece.

"You're basically using your body to short out the circuit to light the LED," says MacCary. "But it's tough getting something to conduct because people are generally really big resistors." The sensitivity of the cloth is so high that the display sometimes flickers in humid weather.

MacCary wasn't always so handy with the soldering iron. Though her father, a former geologist for the government, was constantly building high-voltage devices, she didn't get to spend a lot of time with him. Her parents were divorced by the time she was in high school. Still, she witnessed plenty of the finished projects — Tesla coils, Van de Graaff machines, and Marx generators. "My early experience was getting shocked a lot. He once built a Cockroff-Walton Voltage Multiplier — it's got a lot of capacitors that discharge into it. When you reach the top capacitor, the current is tremendous. It's the sort of thing that would punch a hole in your head if you came too close."

During a recent interactive art exhibit, a local newspaper characterized those in the show as "basement tinkerers" rather than as artists. Several participants took exception, but MacCary says she was pleased. "I don't have a problem with being associated with basement tinkerers. I grew up in that world."

Holter breaks up old sheds in a few minutes thanks to his machine. What started out as a J.I. Case hay baler has undergone extensive modifications and a paint job to become a premium board buster.

Bringing Down the House

Board-busting machine

MAKER	**Cap Holter**	COST	$27 and one rusty hay baler
OCCUPATION	Auto restorer, farmer	TIME	3 days
LOCATION	Glasgow, Montana		

When Cap Holter, 86, describes the origin of the problem, it's hard to get a sense what time frame he's talking about. He says a storm blew down some old sheds on his 1,080-acre property in Montana. A horse barn, hog house, and chicken coop are now a heap of boards. Was it the fallout from Hurricane Katrina? When pressed for specifics, Holter says the buildings were knocked flat while he was in the Army — in 1940.

The Holters — Cap, his two sons and their families — have had these piles of old boards around their homes for years. Now, considering the high price of heating oil, they want to use them for kindling through the winter. But to chop them up would take endless hours with a chainsaw.

During the summer of 2005, Holter put to use the extreme modding skills he learned in the Second World War. The result was an ultrafast board-breaking machine that uses an old two-cylinder hay baler. Now, in under two hours, he and his son Jim can chew through thousands of board feet, filling a one-ton truck and making a huge racket. "We were a little skeptical over whether it was going to work, but it busts those boards just as fast as you can feed them in," says Holter. "The neighbors came by and said, 'What are you people doing up here? It sounds like an automatic rifle.'"

With his son's help, Holter used a blowtorch and an air impact wrench to remove old bolts from a J.I. Case wire-tie hay baler that had grown rusty. They saved the vintage fasteners for winter sorting, and removed the wire-tie mechanism. In a farm shop loaded with welders, air compressors, lathes, and other tools, they took all the extraneous parts off the machine, filling a large truck. "The baler looked like a plucked chicken when we got through," Holter says. He and his son then added a steel guide system on the platform to increase safety. The device works by moving a large reciprocating piston on a crankshaft to shear the boards, much as it used to cut and compact piles of hay into neat bales. "We've been dreaming about this for 30 or 40 years, and we finally got it done," he says.

When talking to Holter on the phone, it's tough to discern his age. He's animated and quick with a story. In the Army, he worked to keep trucks and vehicles operating with whatever means he could find. When he and his fellow soldiers needed an electric welder to do one job, they traded a carton of cigarettes for a Dodge starter generator in northern Australia, and powered it with an engine stolen from an enemy supply dump. "Man, we did a lot of crazy things. Boy, oh boy," he laughs.

Holter was shipped back to the U.S. after catching malaria, and spent the next 43 years working in a body shop. (The farm was always a side venture, since even their substantial acreage doesn't have the scale for a profitable cattle ranch, according to Holter.) Just this year, he restored a two-seat 1940 Packard convertible in the farm shop for a Minnesota customer.

There's always time for odd projects. Holter's son Curt is working on an irrigation system made from an old crane — a swivel on top and a high-powered water jet can irrigate an entire hay field from a single central spot in the pasture. Together the men also hand-built a hydraulic-powered wood splitter. "It sounds like we're all getting lazy here, getting everything mechanized," says Holter. "We'll have to go to exercise class to build our muscles up. Oh, boy."

The indomitable Platypus wins the 2005 Baltimore Kinetic Sculpture Race (top). When not spinning its 30-inch rear paddle wheel, the vehicle glides on truck wheels at up to 30 mph (far left). Bennett (near left) takes a break while stripping the Suzuki Samurai.

The Un-SUV

9-person, 5-speed, pedal-powered vehicle for land and sea

MAKER	**Jason Bennett and David Hess**	COST	$2,400
OCCUPATION	Metalworker, sculptor	TIME	800 hours
LOCATION	Phoenix, Maryland	MORE	kineticbaltimore.com/KSR/2005

Jason Bennett, 26, and David Hess, 41, are veterans of wacky sculpture races — so they knew it was time to get serious about mechanicals. Like other entrants in the Baltimore Kinetic Sculpture Race, they typically made their elaborate people-powered floats out of old bike parts. And yet the vehicles somehow failed during treks over hilly roads, open water, and thick, thick mud.

After Hess and collaborator Karl Saar got an unfortunate 11-foot-high, steel-reinforced Labrador mired during the previous year's competition, the sculptors had had enough. The dog was sold off to a local admirer, and Hess and Bennett immediately trolled junkyards for the backbone of the 2005 entry: a used Suzuki Samurai. "It's the lightest four-wheel-drive vehicle on the market," Bennett says. "We wanted the ultimate driving machine."

Bennett, who was in charge of the functional aspects of the vehicle, incorporated the truck's transmission, transfer case, differentials, drive shafts, tires, and brakes. He extended the drive shaft to 6 feet, shoehorning it to a piece of old steel pipe, and welding that to the transmission's universal joints. He hooked in the transmission and transfer case, and linked the shift controls to the front driver's seat. That way the 4,500-pound, 24-foot-long vehicle, dubbed Platypus, had five regular gears, reverse, and the option of two- or four-wheel drive, high and low. The gear ratio was low enough that one person could pedal it to the starting line, and yet all eight pedalers could hit 30 mph on a hill.

At that kind of speed, it was important to avoid "the meat grinder effect." If the bicycle chains simply turned the drive shaft directly, pedals would start turning on their own with the force of a two-ton mass. To protect the riders' legs, Bennett and Hess heavily modified the power train. At Wal-Mart, they bought eight low-end mountain bikes at $37 each. Bennett cut the rear sprockets out, bought eight larger sprockets, bored them out on a lathe, and TIG-welded them onto the rear cassettes. With the freewheels still attached, riders could pedal at their own pace.

Of course, Bennett had access to some pretty cool equipment. Hess, his partner in kinetic glee, is a professional metal sculptor, specializing in objets of enormous proportion. (He designed, for example, a 6.5-ton steel sculpture hanging from the ceiling of the Baltimore airport.) Bennett has worked in Hess' studio, an old barn at the sculptor's house, and continuously finds new tools. "It's like an unlimited set of Legos really: MIG and TIG welders, plasma cutters, milling machines," says Bennett. "There'll be a vertical metal lathe that he bought at auction that he's never tried out!"

Hess created the look of the Platypus, a vision he describes as "something Devo might drive through Whoville." During the competition, his 10-year-old son, Eli, manned one of the bikes, his wife, Sally, sat in the driver's seat, and his 12-year-old daughter, Sophie, managed the pit crew. In the end, Platypus remained in one piece, won the best art award, and made them think about a much tougher competition in 2006. Says Bennett: "Our secret goal is to kick some butt at the national kinetic sculpture race in California."

A Bug's Life

Insect photography platform

MAKER	**Frans Vandemaele**	COST	€1,500 (about $1,800)
OCCUPATION	Retired engineer	TIME	8 months
LOCATION	Oostrozebeke, Belgium	MORE	users.skynet.be/fotoopa

After four hours aiming at the various hoverflies and bees in his backyard, Frans Vandemaele, 62, found that actually catching them on film was tough. The bugs were always at the edge of the frame, already flying away, or missing from photos entirely. "Only the occasional bumblebee lumbered accidentally into view," he writes in a web journal that's surprisingly riveting. "Other critters flew much too fast. We're not even talking about houseflies, which just refused to be photographed."

Vandemaele, a former electrical engineer, had been fascinated with the creatures ever since he'd seen a museum exhibit. He started his bug-photography device by putting together a precise optical sensor from common laser pointers. As soon as a bug crossed the lightpath, it tripped the shutter. His 13-pound wearable aluminum rack and handmade digital controller jumped into action as soon anything buzzed by. Still, the timing was off somewhere. Bugs were harder to photograph than Sean Penn exiting a club at 2 a.m.

"I had not expected this project would be so difficult," says Vandemaele via email in English. "It was nearly eight months from the start. But I'm retired, so this is my first really free period of time." Vandemaele, whose native language is Dutch, lives with his wife in a cozy suburban neighborhood 30 miles from the North Sea. He started working on the insect photography system in December 2004, just three months after leaving a 26-year job making industrial cameras for giant textile and plastics manufacturing machines.

After its dismal test run, Vandemaele pulled the apparatus back into his home workshop. The problem turned out to be shutter lag, which plagues most digital cameras, especially at the amateur level. When you press the shutter, the camera's electronics need a split second to activate the shutter. Vandemaele's Nikon D100 is a professional-level camera, but it still pauses a tenth of a second.

Vandemaele couldn't afford a new shutter, so he improvised with a 1950s mechanical one he had lying around. On the web, he found a site by a guy in Texas that showed how to hack the shutter and attach a solenoid to fire it electronically. With the help of a friend, he machined an adapter from a 2.4-inch square of aluminum to connect the obsolete shutter to the modern camera body. Now, the hacked shutter sits snugly over the existing one on his Nikon. Vandemaele dials the 21st-century Nikon to stay wide open for 30 seconds, and the 1950s mechanism does the real work of exposing the image — 14 times quicker.

The coding for the project took a month, but the work was relatively easy. Part of the photographer's former job was writing for Altera microcontroller chips, the brains at the heart of many optical industrial automation systems. For the project, he merely repurposed the Altera to shoot pictures of bugs. He uses a freeware app called Quartus for the actual programming, and for mapping out the 98 various input and output pins on the chip to his various shutters, flashes, and optical sensors.

Vandemaele says his toughest challenge wasn't technical so much as zoological. If you hold stock-still in a patch of lilacs for more than 15 minutes, the bugs come to you. They see a guy holding a large aluminum frame festooned with lasers, microcontrollers, and various photographic equipment and — understandably — become curious. "If you have patience, they come to »

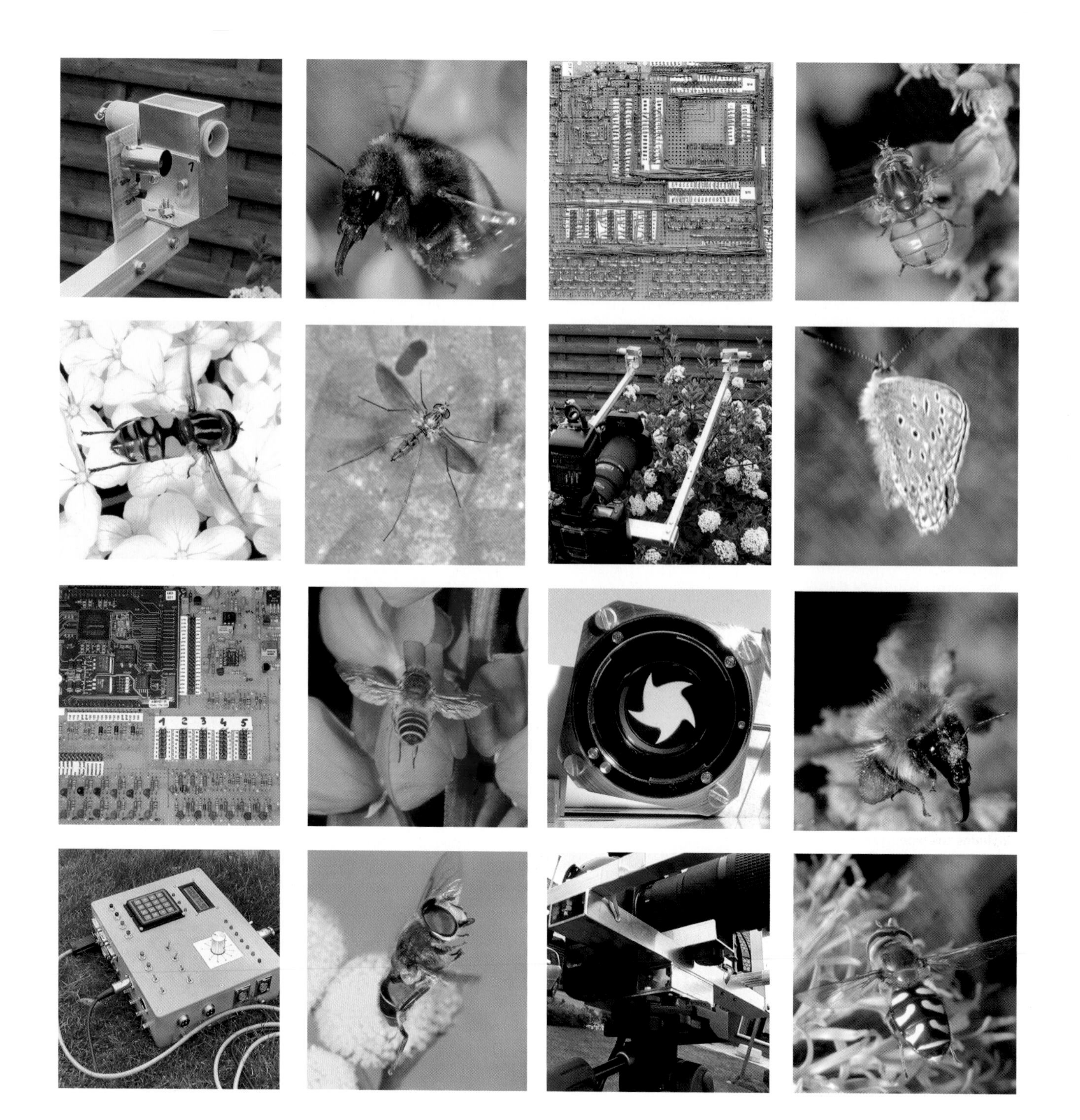
7
1 2 3 4 5

Vandemaele tracks camera-shy insects in the backyard of his home (above). His device consists of a modified Nikon D100 camera mounted on a homemade laser-activated bug detector. Hardware includes the laser pointer and sensor (opposite, top left), custom shutter for the Nikon (third row, third column), and microcontroller box with shoulder strap (bottom left).

» see what is going on there," he writes. "It is just fantastic how these little animals behave, drifting around."

Since the project began running smoothly, Vandemaele has shot thousands of images of insects, and learned a lot. He says most insects fly between 2 and 15 miles per hour, and many flap their wings over 1,000 times per second. Vandemaele says he doesn't need to discover unusual butterfly specimens to hold his interest. The camera gives him a peek into a tiny world, where even ordinary houseflies are fascinating. "It's really funny how flies combat each other," he says. "I once photographed one big fly and one small one. The big one came down and pushed the small one. But then the small one turned right around!"

The builder pushes his electric Porsche 944 down the highway using a trailer he made from the front end of a 1981 Volkswagen Rabbit. The gasoline-powered trailer extends the range of the electric car to more than 300 miles. A view from under the hood (far left) shows an electric throttle control (a series of black boxes on the lower left).

Pedal Pusher

Gasoline-powered trailer that pushes electric cars

MAKER	**J.B. Straubel**	COST	$2,000
OCCUPATION	Chief technical officer	TIME	200–400 hours
LOCATION	Menlo Park, California	MORE	jstraubel.com

With a hankering to show off his custom-designed electric Porsche at car shows around California, J.B. Straubel had a problem. The stretch between Straubel's dorm in Palo Alto and the events was more than 300 miles. His Porsche had earned him a record for electric-vehicle speed in the quarter mile, but like all EVs, its battery only allowed it to travel a relatively short distance. "It gets 30 miles between charges," he says. "I would never make it."

For future trips, Straubel, 29, started brainstorming with friends about a device that would help an electric car bridge long gaps between charging stations. His solution: a gasoline-powered trailer that pushed the car from behind. That way, the electric car could travel long distances, then operate on its battery when it reached its new locale. "When I first talked to friends, they said, 'It'll never work, it'll jackknife, it'll get you killed,'" recalls Straubel.

"I found a VW Bug in a junkyard, sawed it in half, and had it running within a week," he says. With the electric Porsche in neutral, the Volkswagen "pusher trailer" was able to push it up to 60 mph on the freeway. The car's steering and sheer weight made it drive straight despite the force of the trailer, eliminating the danger of jackknife around treacherous corners. "It's amazingly stable," he says. "And you have zero-emission city driving and grid rechargeability wherever you end up."

His latest version is half a VW Rabbit. While riding up front in the Porsche, Straubel controls the trailer remotely using an ignition on/off switch and a small dial for the throttle. The throttle dial is connected to a potentiometer under the dash which sends a signal to a microcontroller on the trailer, adjusting a servomotor and thus the throttle. The Rabbit is always in gear; when Straubel reaches a freeway, he simply starts it up and revs the engine.

If Straubel's friends gave him a hard time, it was nothing compared to the DMV. "I had many, many discussions with them to figure out whether it was classified as a car or a trailer," he says. "There's basically no legal provision for a trailer that can drive itself."

Luckily, he's both patient and passionate when it comes to electric vehicles. "I've been building electric cars since before I could drive," boasts Straubel, a Wisconsin native who revived a dead golf cart from the dump when he was 15, and then moved to an electric bike. When he started as an undergrad at Stanford, Straubel rebuilt the 1984 Porsche 944 into an electric vehicle. He used it to commute to campus through the years while he received his undergraduate degree in mechanical engineering, and a master's in electrical engineering.

The pusher trailer bears a funny little sticker on the back that says "Hybrid." It's a joke, because Straubel hates hybrid cars. He says hybrids are still wedded to gasoline in some way, and pure plug-in cars are the future. "The energy to move an electric car is coming off natural gas, coal, nuclear, hydro, wind, solar," Straubel argues. "It can be a whole bunch of things that are more geopolitically secure, more diverse, and less expensive." He currently works as the CTO of a venture-funded startup company called Tesla Motors, working to create a pure electric car with unprecedented performance. Sounds like if he has his way, we'll all be driving electrics someday — or at least getting pushed down the freeway by personal trailers.

The autonomous aircraft flies at Artists' Television Access in San Francisco (top) and at a Q&A with the creator (left). Source code for flight patterns is on her website.

Home UFOs

Touch-sensitive blimp

MAKER	**Heather Dewey-Hagborg**	COST	$150
OCCUPATION	Student, contract electronics engineer, electronic media artist	TIME	100 hours
LOCATION	New York, New York	MORE	deweyhagborg.com

Fish are supposed to be relaxing, but they're nowhere near as serene as watching your own autonomous dirigible. After a long day, Heather Dewey-Hagborg kicks back with a shiny metallicized blimp, 8 ounces and around 4 feet long, as it dips, spins, and glides around the 20-foot ceilings of her San Francisco warehouse apartment. The movements seem almost sentient. For one thing, she programmed the blimp to be afraid of people, so it scoots away as soon as you touch it.

Dewey-Hagborg, 23, makes her microcontroller-enabled art in her bedroom, so she has some pretty strange equipment next to her nightstand. In one corner, a 4-foot-high helium tank inflates the robot experiments. You'll also see a laptop, multimeter, glue gun, beat-up soldering iron, and her proudest possession, an old 20MHz Hitachi V-212 oscilloscope. "It was something I put off buying for a long time," she says. "Before that, I was making do with the audio input on my PC to analyze circuits."

She wasn't always so close to the machine. In college, Dewey-Hagborg focused on multimedia art. But it wasn't satisfying. "I got interested in lower and lower levels in the technology," she says. "I took a course in programming and taught myself electronics between semesters." On one humid summer day, she went to the RadioShack in the small college town of Bennington, in southern Vermont. "I bought all the classic Forrest Mims III books they had and just went through them every day," she says. "I built 555 timers and op-amps, and then got interested in microcontrollers."

When Dewey-Hagborg moved to the Bay Area, she started a club called SFmicrocontrollers and challenged a friend to a contest of who could build the best flying robot. Her blimp began with a PIC17F876A chip, which she programmed in a C compiler called CCS. The unit has a level switch, so it will shut the motors down when on the ground, and more importantly, a capacitance proximity sensor, which turns the whole aircraft into a single anode with nowhere to discharge until it comes in contact with something conductive, such as human skin. When the component is discharged, the blimp is programmed to move away. A few LEDs blink to show different states of the craft, and three motors help it steer and move back and forth, and up and down. To make the blimp body, Dewey-Hagborg bought silver nylon sheeting by the foot and sealed it with a household iron.

It all sounds pretty labor-intensive, but the artist says she's interested in making lots more flying robots and seeing how they play together. She just started the Interactive Telecommunications Program at NYU in the fall of 2005. "I want to have a whole flock of them," she says. "The robots keep me company."

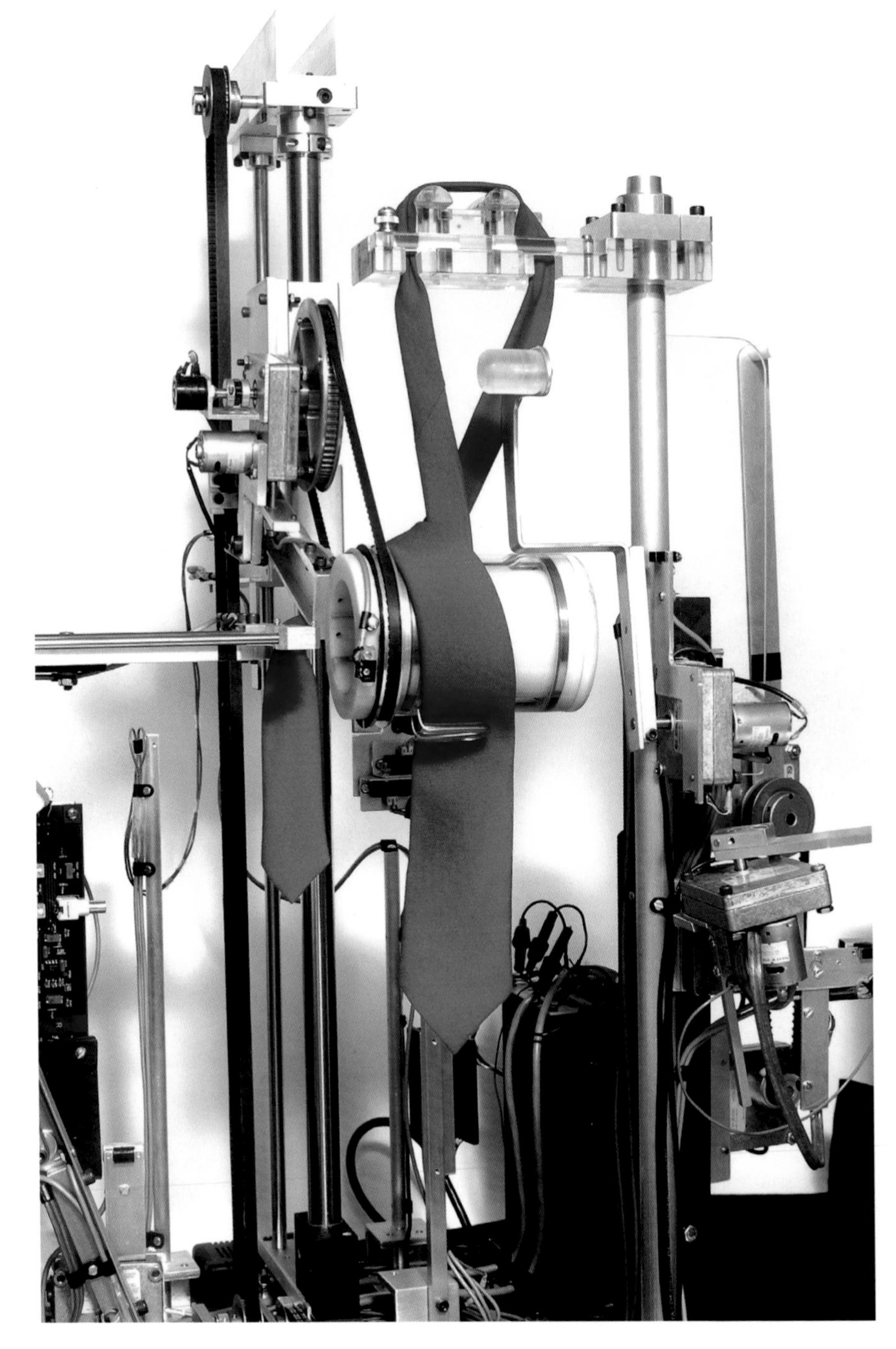

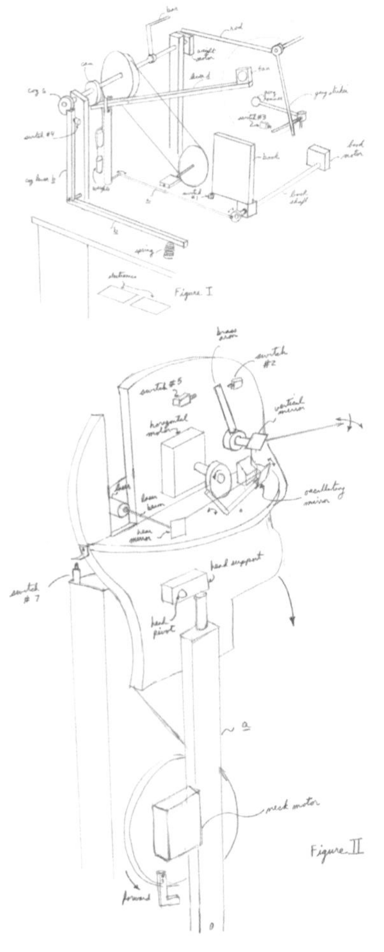

Suave and sophisticated: The new, more efficient *Why Knot* machine (left) ties a four-in-hand in just 350 steps. Goldstein's newest kinetic masterpiece is *Cram Time,* which started out as 3D drawings on paper (above). This view shows the intricate laser path through the figure's eyes.

Why Knot

Necktie-tying machine

MAKER	**Seth Goldstein**	COST	$2,500
OCCUPATION	Retired mechanical engineer	TIME	2½ years
LOCATION	Bethesda, Maryland	MORE	asme.org/education/precollege/whyknot/

And you thought it was a chore to teach a teenager to make a four-in-hand. Seth Goldstein, 65, spent years building a 3-foot-high mechanical device that can prep you for Aunt Ida's family reunion in 562 steps. It ties a knot in nine minutes, unties it, and starts the loop all over again.

Goldstein's wife, Dr. Paula J. Stone, who holds a doctorate in civil engineering, first goaded him to take on the project. He was set to retire in 2002 from a 32-year job at the National Institutes of Health designing biomedical devices. At the breakfast table one morning, she said, "Why don't you make a machine that ties a necktie?" Goldstein cleared out the basement and bought all his favorite shop tools from work: a Smithy 3-in-1, lathe, mill, oscilloscope, soldering irons, and power supplies. During the development, he picked up a rack full of identical red polyester ties as test subjects.

The machine, called *Why Knot,* works by using ten servomotors controlled by separate sample-and-hold circuits. To trigger the moves, a computer program in C runs down a list of steps, sending small voltages to the appropriate circuit to start the motor. Goldstein sighs when he thinks of the effort it took to orchestrate all the analog electronics toward this goal. From the outside, all you see are swinging steel bars and bicycle parts from his old Peugeot.

When word got around that there was a tie-tying robot in his basement, the American Society of Mechanical Engineers chose the device for an exhibit at the Franklin Institute in Philadelphia and gave him a grant for a new, improved tie-tying machine. Because the first version jammed from time to time, he entirely remade the device to use feedback loops, correction routines, and optical sensors coordinated by a sophisticated LabVIEW program on a nearby laptop.

Goldstein, who holds a doctorate in mechanical engineering and a master's in electrical engineering from the Massachusetts Institute of Technology, says he was first excited by machines after visiting a Hershey's factory at age 7. He became absorbed watching them wrap chocolate bars in foil. During his long career, he made such devices as a scanning microscope and a recorder for brain cell signals.

With the attention from *Why Knot,* Goldstein says he'd like to make a go at being a kinetic sculptor. Another monster in his basement is called *Cram Time* — it's a mechanical college student who reads a book as laser beams emanate from his eyes and furiously scan a page. His head slowly lolls asleep, then snaps back up. "A crank lifts up a weight and then this bar comes along and a great big gong goes off and his head snaps back up," Goldstein says gleefully.

One thing you won't hear about *Cram Time* is the word *computer.* Goldstein tries to avoid them whenever possible. As a mechanical engineer, he was dismayed that his partner on *Why Knot,* engineering student Randy Pursley, could write perfect code without ever seeing or touching the machine. With his latest project, everything is coordinated by logic-integrated circuits. "I'm not doing this for maximal utility and efficiency. That's one of the advantages of retiring," he says, pointing out that even his planning process is on paper. "People say I should be doing the drawings on CAD, but that's not fun. To hell with it."

The magnetometer, delicately balanced between two coils (above), reflects changes in the Earth's magnetic shell 40,000 miles away. At the heart of the device, a magnet (left) sits on a plexiglass shard. The tiny mirror glued behind it reflects a laser beam for fine adjustments, and the silver half-dollar provides dampening.

Backyard-o-sphere

Web-connected magnetometer

MAKER	**Joseph DiVerdi**	COST	$100
OCCUPATION	Scientist	TIME	2 years
LOCATION	Fort Collins, Colorado	MORE	xtrsystems.com/magnetometer

The Earth wears a huge protective shroud of magnetic energy that looks like something like a kid's ghost costume at Halloween. The narrowest part extends toward the sun 40,000 miles, while the tail hangs into deep space for millions of miles. It's called the magnetosphere, and it generally works as a shield for deadly particles from solar winds and coronal mass ejections. Scientists believe this big buffer helped us hold onto our atmosphere during the Earth's formation.

For a little over a year, Joseph DiVerdi's computer checked in with our magnetosphere once a second. DiVerdi, 51, built a super-sensitive device called a magnetometer, which could detect minor fluctuations all those miles away, and he buried this box of equipment under a blue spruce tree near his garden, 30 yards from his house. A serial cable shuttled data back to the computer in his home office.

DiVerdi, who holds a Ph.D. in chemistry and has worked as a web developer and trainer, jump-started his project when he read about a simple homemade magnetometer made by a fellow amateur in Texas. He immediately followed the plans, balancing two small magnets and a mirror on a taut nylon fiber. He positioned a separate magnet to cancel out the effects of the Earth's natural magnetic pole and shined a cheap laser pointer on the mirror. Then, incredibly, minute changes in the Earth's field showed up in the laser beam. DiVerdi one-upped the inventor by rigging an electronic sensor to capture the movements. A Perl script sent the data to a PC server running Apache.

The goal? To give away hard scientific research for anyone to use. "People need to create more data, whatever their area of expertise," says DiVerdi. "Just slurp it up with an old PC card and slap it online."

The graphs began to accumulate on DiVerdi's site, but there were some strange readings. It turns out the magnetometer was so sensitive that it was getting thrown off by people moving chairs or carrying metal tools around the house. To give it a home away from home, DiVerdi created a 4-by-3-foot box out of ¼-inch plywood, and sited it in the backyard.

DiVerdi — who lives with his wife, a professor at Colorado State — has completed more mundane projects like potato cannons with his grown son and daughter. But his latest amateur venture also deals with outer space — a radio-based system for measuring changes in the Earth's ionosphere. Despite his idealism about reaching out to other scientists with his findings, the magnetosphere project has already taught him that scientific communities are fairly insular. "I'm not one of the in-crowd, so my data is routinely ignored," he says.

It's too bad, since projects like the garden magnetometer could do such good. After all, ever since the magnetosphere was confirmed by satellite in 1958, scientists have been looking for ways to tie solar events with terrestrial weather. More recently, the owners of deployed satellites and human space missions have sought timetables for when harmful solar particles are on the rise. DiVerdi's graphs show the effects of daily solar events as well as seasonal rhythms in the magnetosphere. Maybe no one can believe that something so valuable could really be free.

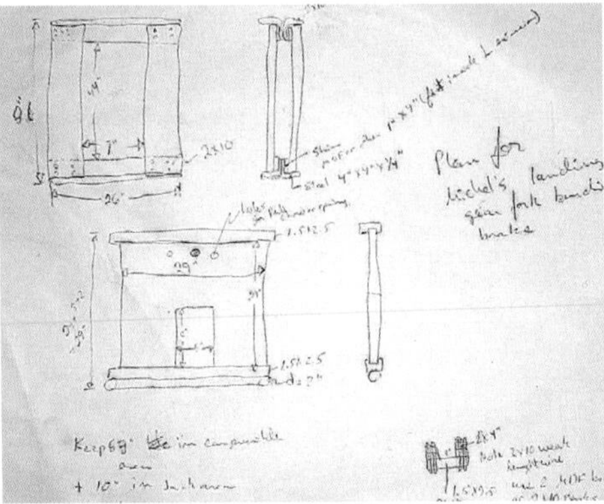

Therrien hand-built his Zodiac kit aircraft (top) from a bag full of parts. To create its curved ribs and wings, the DIY fundamentalist went so far as to design his own hydraulic press (left) from 2-by-8 boards and a car jack.

Flying from Scratch

Homemade 6-ton hydraulic press

MAKER	**Michel Therrien**	COST	$60 or less
OCCUPATION	Management consultant	TIME	4–6 hours
LOCATION	St. Julie, Québec	MORE	mthobby.pcperfect.com/ch601/press.htm

Michel Therrien went into an aircraft machine shop looking to bend five pieces of aluminum. "They wanted to charge me $200. It was about 15 minutes of work. It didn't make any sense," says Therrien, who recently finished an airplane from a kit that took six years of work in his spare time. His Zodiac CH 601 HD once consisted of a $13,000 plastic bag full of parts and a grimy 1983 Subaru Legacy engine he'd bought cheap from a scrap yard.

What you need to know about amateur aircraft builders is that they aren't just thrifty, they're radically DIY. By aviation law, you must build 51% of the plane to fly it as an "amateur-built" aircraft. By tradition, gonzo builders do nearly 100%. Instead of patronizing the machine shop, Therrien, the 39-year-old boss of a management consultancy in Québec, Canada, decided to recreate several exotic tools himself to finish off his plane.

To make his own metal-bending machine, Therrien stopped off at Home Depot to pick up a $25 hydraulic jack, 2-by-8 boards, and ½-inch bolts. He built a central wooden bar beneath the jack that would create downward pressure and an outside form as a brace for the jack to push against. By manipulating the wooden surfaces and exerting 10 tons of pressure in a small area, Therrien's hydraulic press bent a ¼-inch aluminum plate at just the right angle to complete his landing gear. And with the success of the prototype, he made two more versions to create flanged holes for rivets in the airplane's ribs and to bend the leading edge of the wings' skin. Each press took about five hours and cost about $60 to make.

The 1,000-pound Zodiac, with its 30-foot wingspan and 135 mph top speed, was finished in Therrien's single-car garage by mid-2004, but wasn't licensed to fly. A friend certified the craft, and then Therrien used it to get lessons, receiving his license in just three months. Soon after, his wife agreed to fly with him on Father's Day, leaving their two sons on the ground. He admits she was a bit leery, but insists that she was worried about flying in general. "She wasn't once worried about the craftsmanship."

Brainy Bucket

LED-enhanced bicycle helmet

MAKER	**Ted Selker**	COST	$40
OCCUPATION	College professor	TIME	1 day
LOCATION	Arlington, Massachusetts	MORE	web.media.mit.edu/~selker

If you see a bicyclist wearing an LED-adorned motorcycle helmet in Boston traffic, steer clear. It's Ted Selker, 49, an MIT professor who spent the last 20 years on a personal quest for the perfect mix of security and entertainment on his mountain bike.

For one recent headgear setup, Selker modded a bright red Shoei motorcycle helmet he found at a garage sale. In back, horizontal strips of LEDs signal the cyclist's left and right turns. (Switches for each direction are near Selker's coat sleeves.) Inside, one earbud plays a book on tape, the other connects to his cellphone. A personal protection siren, bought at a drugstore, dangles from his chin strap for blasting negligent drivers.

Selker drilled holes in the mouthpiece of the helmet, then used a glue gun to embed more LEDs. "According to the psychology of drivers, I came to realize that they pay more attention to me in traffic if the warning lights are positioned in my mouth," says Selker, who commutes 12 miles round trip from Arlington to Cambridge. (Friends say the headgear's downside is that he doesn't bother to take it off to chat.)

In 2005, Selker's most advanced helmet yet came out of a collaboration with his grad students. To help the rider mediate between the sounds inside and outside, the helmet turns music off automatically when you hear a car horn. If the helmet hears you shout at another driver, a 130-decibel alarm sounds. When the bicyclist wants to signal a turn, a nod of the head triggers the left and right blinkers. (The electronics obey gestures thanks to an accelerometer and PIC microcontroller.) Pretty impressive. But not as fun as Selker's low-tech headgear, personally tested down Massachusetts Avenue.

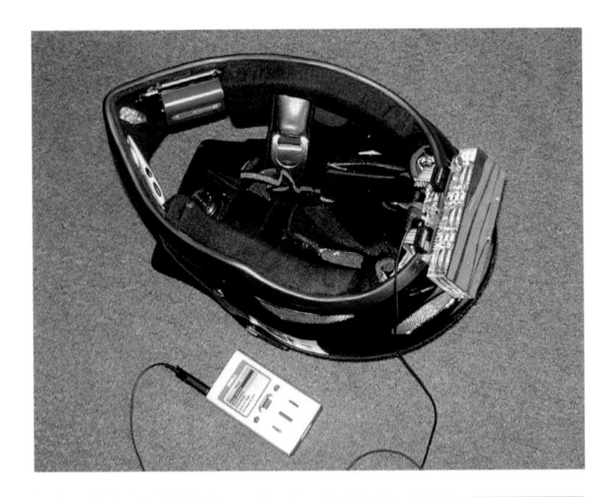

Think of it as a dashboard for your head. The media helmet illuminates left and right LED blinkers whenever you purposefully nod in one direction. And it will squelch the music player or cellphone if it hears a loud sound on the street, forcing you to pay more attention.

20-Horsepower Nut Grabber

Chestnut harvesting device

MAKER	**George Turnbull**	COST	A$25,000 (US$18,900)
OCCUPATION	Farmer	TIME	8 years
LOCATION	Stanley, Australia		

The Australian chestnut season starts in late March as soon as the nuts fall on the ground. George Turnbull once had 14 pickers struggling to grab 10 tons of nuts from 500 trees, but then the bottom fell out of the chestnut market. He could no longer afford the help. In a 10,000-square-foot barn packed with junkyard gears, welding machines, and hand tools, Turnbull, 68, started to cobble together a machine that he hoped would change the industry.

Part ride-on mower, part street sweeper, and part combine harvester, the Turnbull Multinut Harvester scoops up nuts off the ground and stores them in a bin behind the driver. Yellow brushes in front sweep the nuts away from the wheels to avoid running them over. Two stainless-steel shoes underneath follow the contour of the ground, moving up and down a total of 1.5 inches, and sucking up nuts into an elevator inside. Turnbull, a former miner, truck driver, and timber puller, enlisted his son to help him make new prototypes. "Every year, I think, 'I could do better than that,' so we pull it back into the shed," says Turnbull.

Currently, they are on version seven, which uses hydraulics (made by a third party) to streamline the machine. Now all three wheels are driven by hydraulic motors. A little joystick beside the driver steers the back wheel. It can turn tightly around the trunk of the tree.

Turnbull also invented a mobile deburring machine to pull off the hard exterior shell from the nut before it's graded and bagged. The nuts still have the familiar leathery brown skin on the outside, which still must be hand-peeled by workers in China. Says Turnbull, a maker to the core: "I've got a lathe coming in a few weeks ... I'll find a technological solution to get 'em peeled too."

The Multinut Harvester is the result of long hours of welding, machining, and hunting through local junkyards. The latest, "Mark 7" machine, pictured here, uses a hydraulic motor drive to keep the collector low to the ground for passes under tree branches.

In Sandy Hook, N.J., the radio-head's first van (top) picks up signals from hams across a wide band of frequencies. In summer 2005, he upgraded his snazzy Honda Element (left) with a 4-foot microwave dish to catch ultra-high-frequency waves.

Microwave Ranger

Ultra-high-frequency rover van

MAKER	**Sigurd Kimpel**	COST	$1,500, much of it donated
OCCUPATION	Retired engineer	TIME	60 hours
LOCATION	Pittsfield, Massachusetts	MORE	mgef.org

In the world of amateur radio, a few tasteful whip antennas on top of your car can be a totem of power and competence. Sigurd Kimpel's new Honda Element — with its 4-foot round microwave dish and multiple aluminum Yagis — gives him street cred to spare.

Kimpel is part of the Mount Greylock Expeditionary Force radio club, and his mobile tranceiving station — a rover, in radio parlance — allows him to conduct radio experiments by driving to an area of best reception. "My rover is by no means the most impressive, but it is one of the most compact and it's great on gas," says Kimpel, who's better known to friends by call numbers KJ1K.

After years of working in the more typical HF bands of amateur radio, the 66-year-old retiree was looking for a bigger technical challenge. After all, there are public radio frequencies dotted along the spectrum all the way from 1.8MHz to 300GHz. Most hams conduct communication under 1GHz, and are able to talk across the world. The higher frequencies have much shorter ranges, but are a tougher technical challenge. One local hero to Kimpel, for example, talked along the 271GHz band. The radio energy was nudging close to the visible light spectrum, and only allowed him to talk a few hundred yards. The longest recorded range for Kimpel's van is 260 miles, from Mount Greylock, Mass. to Cape May, N.J.

Microwave-band equipment has to be homemade. Kimpel takes off-the-shelf ham radios and mods them with transverter circuits, which raises the frequency many hundreds of megahertz. He designs the boards himself and trades designs with others in the club. In addition to the homebrewed transceivers, Kimpel has a laptop for refining signals digitally. A 90-amp-hour marine battery and a 1,000-watt Honda generator power it all (the Element's battery wouldn't stand a chance).

When Kimpel showed up at the Honda lot in early 2005, he liked that the Element was small and had a rear sunroof for snaking antenna feed lines. And according to his detailed wind-force calculations, the exterior was sturdy enough to hold the dish and individual low-band antennas. "I'm comfortable running 65 mph on the freeway," he says. "Well ... unless a motorcyclist is tailgating me."

"On a typical weekend, I leave on Friday night and I come home Sunday night, and I put between 800 and 1,000 miles on the vehicle," says Kimpel. He'll drive as far as Huff Corner, Maine, and bring friends to compete in national contests to log the greatest number of successful radio contacts. His wife, also a ham, usually stays home from these junk-food-fueled jaunts. "We don't get much sleep," he says. "I'm usually busy trying to help the club win."

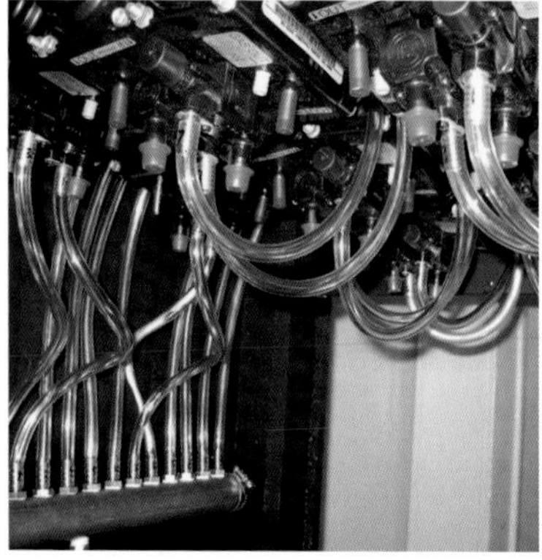

Wollborg, Baumstark, and Strech (top, left to right) sample cocktails from a menu of 200 on their original PC-based bartender. While their first version used 16 valves, a new bar-bot employs 24 electronically controlled CO_2 gas valves (left). The result? Over 1,000 tasty combinations.

Cocktails at 700 Mhz

Touch-screen electronic bartender

MAKER	**Nate Strech, Joel Wollborg, and Torrey Baumstark**	COST	$1,300
OCCUPATION	IT systems engineer, app/mech engineer, and IT systems analyst	TIME	3 months
LOCATION	Ann Arbor, Michigan	MORE	aibar.net

Think of it like having R2-D2 as your own personal bartender. You can talk to your friends during parties without being chained to the bar, and impress them with hundreds of exotic cocktails listed on a touch-screen — from a Bat Bite to a Singapore Sling.

That's the vision that enticed three University of Michigan roommates to build their own electronic, PC-powered Sam Malone. After looking around the internet for design suggestions from other home bar engineers, Nate Strech, Joel Wollborg, and Torrey Baumstark started with an old wooden TV console donated by a family member.

They set up their machine to use 16 different liquids, including things like tonic water, orange juice, sour mix, and vodka. Finance major Strech tested his Visual Basic programming skills to link a cocktail database in Microsoft Access to a relay controller built by electrical engineering major Wollborg. For each bottle, a 5-amp relay activates a valve that shoots a colorful mixture into a guest's cocktail glass through a chrome pipe at the top of the machine. All 16 tubes run inside the pipe so liquids don't mix until they are supposed to. The most expensive parts of the project turned out to be 16 air-driven beverage pumps from specialty beverage supplier Flojet. They're powered by a CO_2 tank nestled inside the cabinet.

It worked, but all was not champagne and caviar. Sodas went flat, juice pulp plugged the valves, and people missed the look of classic liquor bottles lined up on top of the bar. Two years later, the roommates have graduated and have engineering and computer jobs around Ann Arbor. They still live together, and are still searching for the perfect bar. Their latest prototype uses airtight rubber seals for bottles so they can be permanently displayed above the bar. Mixers stay fresh in an accompanying mini-fridge, and the whole business is powered by pressurized CO_2, obviating the need for pumps.

What's the report back from the lab? "We're hopeful about the viability of our new theory," says Strech. "We're just finishing up a lot of testing."

Cub Scouts from all over Southern California compete at The Oaks mall in Thousand Oaks (above). A chip in the 96-foot Pinewood Derby track automatically coordinates race results on the digital projector. Scouts control the starting gates (left) using buttons at the finish line.

Against the Grain

Automated Pinewood Derby track

MAKER	**Paul Taylor**	COST	$6,000
OCCUPATION	Communications equipment installer	TIME	Many days over 9 years, constantly upgrading
LOCATION	Westlake Village, California	MORE	hometown.aol.com/pwdtrackmaster/track.html

A heady combination of NASCAR and Norman Rockwell, the Pinewood Derby consists of hand-carved wooden cars hurtling down a runway to the roar of hundreds of Cub Scouts and Scoutmasters. The competition has been run in gymnasiums across America for almost 50 years. So it's not surprising that when Pinewood dad Paul Taylor proposed major changes in the traditional track, he instantly drew criticism from longtime scout leaders.

Taylor, 58, a communications installer for Lucent, noticed that the boys who build the cars aren't allowed to race them. In many derbies even today, an adult loads the cars, calls the start, and releases the models all at once. "That's a piss-poor way to do it," says Taylor, telling the story of how his son Brandon, then 8, sat on the sidelines as his two chances at race day came and went. "You get left out and lose interest."

Taylor immediately retreated to his garage and cobbled together a new type of track that would allow the scouts to start their own races. In his vision, the boys themselves would hover over big red buttons as drag-strip style lights count down the start. A weight sensor makes sure the cars are loaded properly, counts reaction times, and trips a red "default" light when a scout jumps the gun. Infrared sensors time racers to within 0.001 seconds, automatically tallying winners. The new system would mean no adults bickering over who won, and, with the time saved, giving scouts at least four tries.

With the prototype track loaded into his Isuzu Rodeo, Taylor made a presentation to scout elders, who summarily rejected it. A few months later, a local pack assented to try the new system. That was in 1996.

Today, Taylor's track is the standard for over 20 packs and two district races throughout Southern California. Over the years, Taylor has added pre-recorded voice samples that announce the winners and an 8-foot projector screen that shows digital photos of the children's faces as they race. The track now takes center stage at shopping-mall-based events and scout conferences.

Taylor's track runs on Microsoft Excel, of all things. Infrared sensors and pressure switches feed into a 6502 microprocessor from a vintage Commodore 64 computer. The chip sends keystrokes through an IBM wireless keyboard to a nearby laptop. The keystrokes automatically enter competition times in a large Excel spreadsheet and trigger Excel macros to coordinate sounds and images. Taylor programmed the intricate scripts himself in the desktop accounting software.

"Before a race, I take the kids' pictures with a digital camera and enter their names. Once we start, all I have to do is sit back," says Taylor. Race statistics show up on an InFocus DLP projector as Taylor's prerecorded voice booms on a set of speakers. If a car crashes on the track, a huge explosion sounds and flames appear on the screen. Taylor's disembodied voice announces: "CAR NUMBER — 33! — HAS BLOWN ITS ENGINE. FORTUNATELY, THE DRIVER WILL WALK AWAY FROM THIS ONE."

Other scout troops around the country have built optical timing systems, but Taylor's is the only one that uses scouts' reaction times to start the races. And his has the benefit of his announcements to spur young racers. "This track has got it all," says Taylor. "Who else is going to put nine years into something like this?"

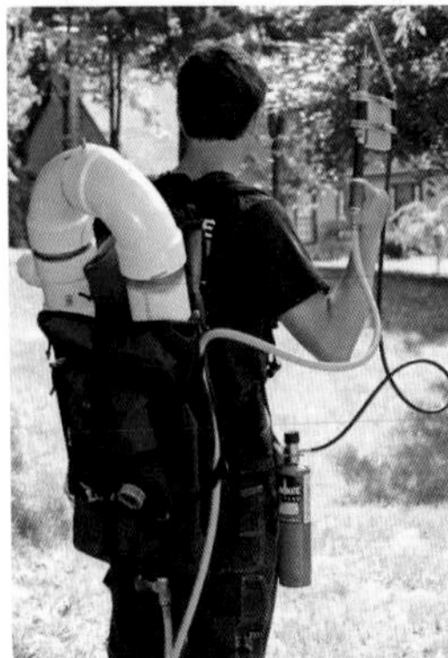

Monaco lights up the cul-de-sac of his Pennsylvania neighborhood. The chemical engineering student created his personal torch using denatured alcohol and a backpack-mounted pressure tank.

The Long Tail

Flamethrower from plumbing pipe

MAKER	**James Monaco**	COST	$250
OCCUPATION	College student	TIME	1 week
LOCATION	Kingston, Pennsylvania	MORE	procrastinet.com/archives/000064.html

It was an intimidating online persona, to say the least. Pictured with a tank of liquid fuel strapped to his back, the pseudonymous "Roland Tower" began his online project description: "If heading on down to the local Home Depot and transforming $100 worth of random pipe bits into a killing machine doesn't appeal to you, you're a frikkin' pansy."

With that, the members of various web forums immediately began speculating about the identity of this suburban commando. Was he from a broken home? Was he capable of a Columbine-like act of terror?

No. It turns out that Roland Tower is really James Monaco, 20, of Kingston, Pa., and a friendly, well-adjusted young man. The Penn State sophomore in chemical engineering swims up to two hours a day for exercise. He got the idea for the flamethrower after creating a pneumatic spud gun the year before. The same air pressure that launched vegetables, he figured, could spray water from a plumber's valve. "At first I thought I was going to make a really kick-ass Super Soaker," he says.

Monaco gathered the parts: a 6-foot hose, 4-inch-diameter PVC pipe, Schrader valve like those used on bike tires, and bicycle pump to pressurize the tank. He created the water tank in his driveway with a reciprocating saw and a few cans of PVC cement. An old camping backpack would hold the apparatus while he worked the nozzle.

Then, inspiration struck: "I thought, 'What if I could use denatured alcohol instead of water?'" says Monaco. He knew gasoline would eat right through the plastic pipe, but not alcohol. Instead of the bike pump, he would carry a small bottle of nitrogen from a welding store. That way, there would be no possibility that fire would creep back up the hose and ignite the backpack.

"My parents thought it was a water gun right up until everything was glued together," says Monaco sheepishly. "Then I casually brought up the fact that I could put alcohol in it. They reacted very poorly at first."

After talking it through a little, however, Monaco gradually reassured his parents that the system would be safe for controlled experiments. (Monaco's father is a pediatric cardiologist at a small group practice in northeastern Pennsylvania. His mother, also an M.D. by training, is a homemaker.) Ground rules were set. He would never run the device without the supervision of one parent standing by with a fire extinguisher.

The first test went very well, with Monaco's mother both proud and terrified. "We took some video where you can hear my mom alternating between talking about how amazing it is and screaming, 'Get away from the trees!'" he says.

Monaco's interest in science drives him to experiment both in and out of school. For his summer job, he prepped research equipment at the university lab. In his college dorm, he and his roommates created a liquor still from plumbing parts and an old teakettle.

But it's the flamethrower that has drawn the most interest from those in his neighborhood. Once, while Monaco blazed a 20-foot geyser of flame straight into the air, he heard a small girl squeal with delight. Her father, a former pharmacist, walked quickly behind her, and Monaco turned off the system and braced for a lecture. "But he was absolutely thrilled with it," says Monaco. "He started giving me advice on how to make an exploding hydrogen balloon with a magnesium wick."

One of the toughest challenges of building a super-ATV around a burly Subaru power plant was fitting the steering and suspension around the cramped engine, says the maker.

Speed Demon

ATV from salvaged Subaru engine

MAKER	**Ken Brough**	COST	$40,000 plus sponsored parts
OCCUPATION	Vehicle inspector, construction worker	TIME	2,000 hours
LOCATION	Te Awamutu, New Zealand	MORE	stiatv.co.nz

Ken Brough, 50, doesn't talk much about the crazy rides he's taken on his home-built ATV. One gets the sense that he's in an odd predicament. His radical four-wheeler isn't street legal according to New Zealand government officials, so he tries to pretend it doesn't rip up the backroads of his tiny hometown, 100 miles south of Auckland.

To be sure, this quad is a little hard to categorize. It runs on a second-hand 2500cc engine from Subaru's fastest car, the WRX-STi. But it also has handlebars, a gearshift pedal, and a suspension like a trail-burning dirt bike. Most ATVs don't have fancy limited-slip differentials, but Brough's all-wheel-drive Subaru transmission has three, giving him unheard-of speed and traction on sand dunes and dirt paths.

"They were scared off when I presented it," says Brough, recalling his arguments with the Land Transport, the office in charge of registering vehicles. "They threw up their arms and said, 'If it's a car, where are the seatbelts?'"

Brough designed the ATV not for roads, but as an entry in a local hill-climb race. He convinced a friend and Subaru parts dealer to sponsor him for the engine, then built the frame out of 38- and 25-millimeter roll-cage tubing. Instead of a radiator, the vehicle circulates water for cooling through these sealed frame tubes. Motorcycle handlebars connect to a hacked steering mechanism shoehorned under the engine and to one side of a modified suspension. All five gears and reverse shift the same as the H-pattern in a car, except that the gearshift is a small wire stirrup that goes around the driver's left foot.

A trained auto mechanic, Brough currently works part time in construction. Coincidentally, he also has a job two days a week as a vehicle inspector for Land Transport, the agency that won't register his quad. Whether he's working on cars professionally or not, he's always tinkering with a Datsun 240Z or some new creation in the one-car garage behind his house. Ten years ago, Brough made waves with a new type of speedway midget car, which won a few midget series races, but was soon disqualified by competition officials. His two daughters from a previous marriage, 18 and 21, bring friends and boyfriends for peeks at his latest R&D.

As soon as the Subaru ATV was done, Brough couldn't wait to race it uphill. "The power-to-weight ratio is awesome," he says, admitting to a helmetless quickie in which he hit 85 mph. All the horsepower of a sports car is packed into a tiny, 1,168-pound package. Unfortunately, race officials at the uphill climb immediately disqualified the Subaru ATV. Brough was crushed.

Perhaps vindication came at a recent auto show, where Brough's quad received accolades from professional race-car drivers as well as ordinary "petrol-heads." "People came up to me saying, 'It's a work of art,'" he says. "But I was thinking, 'It's just a friggin quadbike I built.' While everyone else is dribbling in the mouth, all I can see is a ripple in the chrome work." The debut also fostered dreams of bringing the ATV to the U.S., where he suspects the vehicle registry is more lenient. At the rally, one man took video footage of the quad on its pedestal. "I was suddenly looking at it in a different light, and I said, 'My God, that's something that should be in Las Vegas!'"

Mark

Hardwood Jets

Wood-powered turbine engine

MAKER	**Mark Nye**	COST	Can$6,000 (US$5,120)
OCCUPATION	Machine shop owner	TIME	200–300 hours
LOCATION	Mississauga, Ontario	MORE	nyethermodynamics.com

You could say Mark Nye has some hands-on experience with jets. He attached one to the family boat, connected one to a car, and loves firing them up at work to blow off steam. But his latest gas turbine has a bigger goal than pure fun — it could just be the key to cheap, alternative energy in rural settings.

Nye's breakthrough jet burns wood, a commodity he has plenty of. The 43-year-old lives in Mississauga, Ontario, where he runs a busy welding shop filled with the stuff. The 40-person company makes specialty parts for heavy equipment. But the huge steel sheets they use arrive in wooden crates called dunnage, which often goes to waste. "I sat and thought, is there a way to get rid of this stuff and make electricity?" says Nye.

One day, he came up with a solution. He'd been building jet engines since 1985, often using rusty turbochargers from cars. An automotive turbocharger's quickly spinning fan will compress fuel just like a turbine fan inside a real jet engine. Nye has a bunch of turbos crowding the floor near his desk nearly all the time. ("Don't ask me why," he says. "To tell you the truth, I just like looking at them; they kind of turn me on.")

He welded the turbo from a mid-1980s Dodge 2.2-liter truck engine to a reinforced steel barrel and tossed in finely chopped cordwood. Air gets sucked into one end and compressed by a fan spinning at up to 140,000 rpm. The quickly moving fresh air stokes the fire. A 1,100-degree Fahrenheit plume comes out the other end, turning another fan. The two fans are connected — the faster the fire burns, the more air is sent in to further stoke it. Nye says the steadily increasing whine is deafening — and exciting. "Nothing will compare to the first time I ran it. The bloody thing immediately worked with no tweaking. There's nothing as glorious as that."

He soon created a second-generation wood jet (pictured at left) with a more sophisticated air circulation system. After a few adjustments, the machine turned logs into a white-hot stream of gases with a compressor discharge pressure of 20 psi. "There was so much pressure it blew the hose out of the manifold. It scared the crap out of us." Nye estimates that it's enough mechanical force to generate as much as 2,000 watts of electricity — enough power for a small, rural business.

The builder says his intense curiosity about jet engines started at the Toronto International Dragway years ago. "We went to the drags when I was a young boy and saw the Green Mamba, Earthquake, and all the other jet-powered cars," recalls Nye. "I was just blown away."

His fabrication shop is the perfect space to stage advanced jet R&D. It's filled with computer-controlled machines for automated milling, turning, and sawing. Nye and his colleagues practice heavy-duty flux core welding of steel up to 4 inches thick (one of their huge crane attachments was used to drag steel beams out of Ground Zero after 9/11). Nye, who earned a two-year mechanical engineering degree in college, inherited the business from his father and works there full time.

After making jets as a hobby, his projects started paying for themselves. A University of California, Berkeley professor bought a jet engine to use in the classroom, and an early jet boat sold to an enthusiast. With the money, Nye created *Squirt 2,* a boat that used a surplus U.S. Air Force helicopter engine. It runs on diesel, biodiesel, stove oil, and jet fuel (which he gets at the airport for $2.75 per gallon), and he takes it out with his three kids — ages 7 to 13 — for cruises on Lake Ontario. »

"I'm afraid to floor it," says Mark Nye as he lights the afterburner of his jet boat, *Squirt 2.* Thanks to a surplus U.S. Air Force helicopter jet engine, the craft has 1,400 horsepower for a total of 99 mph. "It's just an immense amount of power," says Nye. "It really scares my wife." Opposite page: *Squirt 2* in action, along with the jet fanatic's hand-built experimental engines. His dragster design started out as a turbine mounted to the front axle of a Pontiac 6000 LE (top left). His afterburner model (top row, third from left) sold to a university in California.

» The boats and homebrew dragsters may be loud, fun, and conspicuous, but Nye's heart resides lately in his serious research. He described his mood the chilly January day the wood jet sustained a continuous burn using premium cordwood. He stood over the round combustor like it was a big cauldron, and the only problems he saw were "clinkers" — small, hard pieces of fused ash that can wear out a turbo. Otherwise, everything went as planned, and Nye celebrated the auspicious event over Molsons. "I'm Canadian!" he shrugs.

Cover photography by Fred Metoyer and April Guy

8–10, illustrations by Damien Scogin

16–17, illustration by Sara Huston

18, *Doughnut R&D*: photograph by Lee Flamard

20, *The Life Subaquatic*: photography by Claus Nørregaard

22–25, *Space Camp*: photography by Dean Carothers and Bess Newton

26, *The Power of Light*: photograph by Marcel Betrisey

28–31, *Cool Worlds*: photography and illustrations by Bathsheba Grossman

32, *Tabby Tracker*: photograph by Bob Blick

34, *Intersection*: illustration by Rockwell Schrock

36–39, *Outside the Box*: photography courtesy Tom Chudleigh

40, *The Sound of Money*: photography by Thiago Teixeira

42, *Scale Model*: photography by Koichi Hirata

44, *Infra Rad*: top, photograph by Gillian Zaharias; bottom, photograph by Greg Miller

46, *DJ Skunk*: photography by PeCan

48, *Best Part of Waking Up*: photography by Rich Hauck

50, *Fake Mac*: photography by Hideo Takano

52, *Live and Let Fry*: photography by Thomas Schnackenberg and Louis Giersch

54, *Pick of the Litter*: photography by SorgoNet Team

56, *Living on a Thin Line*: photograph by Owen White

58, *Art Machine*: photograph by Anton Perich

60, *Horse with No Name*: photograph by Douglas Repetto

62, *Horse with No Name*: top and bottom left, photography by Douglas Repetto; bottom right, photograph by Terrence Pender

63, *Horse with No Name*: photograph by Douglas Repetto

64, *Mr. Punch*: photography by Volker Morawe and Tilman Reiff

66, *The Cuisinart Whisperer*: photograph by Hyun-Yeul Lee

68, *Parts Mart*: photograph by Zach Radding

70, *Private Jet*: photograph by Dr. Diana Krohn

72–73, *Private Jet*: photograph by Hans-Joerg Krohn

74, *Holy Workspace!*: photography by Anab Jain and Stuart Wood; illustration by Sohui Won

76, *Waterbug*: photograph by Chris Lebakken

78, *Chitty, Chitty—Bang! Bang!*: photograph by Richard Flanagan

79, *Ersatz Elevator*: photograph by Jack Buffington

80, *Cool Shirt*: photograph by Wendy Tremaine

82–85, *Two Wheels Good*: photography by David Anderson

86, *Kite Aerial Poaching*: photography by Forrest M. Mims

88, *Batman*: top, photograph by Bianca Hester; bottom, photograph by Scott Mitchell

90, *Shoot Out the Lights*: photograph by Peter Ubriaco

92, *Tangled Up in Blue*: photography by Reagen Ward

94, *A Real Mashup*: top, photograph by Frank Sanns Jr.; bottom, photograph by Richard Hull

96, *Blinding Technology*: photograph by Sharon Drake

98, *Shocking Developments*: photograph by Jennifer Hubbard

100, *Catch and Release 1.0*: illustration by Neil Fraser

102, *Retro Chic*: photograph by Terry Nordbrock

104–105, *Retro Chic*: photography by David Forbes

106, *Fetch, Rover*: photograph by Fred Metoyer

108, *Super Scope*: photograph courtesy Janet Davis

110, *All Mod Cons*: photography by Ryan O'Horo

112, *Surrender Dorothy*: photography by Harald Edens

114, *Java Development Environment*: photography by Frank Voelkel

116, *Block Head*: photography by J.P. Brown
118, *Pooch Power*: illustration by Tim Hunkin
120, *Lionel Hampton on Autopilot*: photography by Larry Cotton
122, *PVC Piper*: photograph by Dennis Havlena
124, *Surf and Turf*: photograph by Veronica Harrison
125, *Run Silent, Run Deep*: photograph by Philipp Rott
126, *Cheap Heats*: photograph by Justin Yarrow
128, *Road to Nowhere*: photography by Yuval Tal and Zvika Netter
130, *Industrial-Age Origami*: illustration by Ed Bertschy
132, *Sun Hog*: photograph by Chris Dunklee
134, *The Speed of Starch*: photograph by Andy Gustafson
135, *Six Monkeys*: illustration by Al Gori
136, *Ace Service*: photograph by Dimitri Merrill
138, *Apparition Apparatus*: photography courtesy Alexander Telegin
140, *Radio Free Boston*: top, photograph by Vincent Leclerc;
bottom, photograph by Limor Fried
142, *Soul of an Old Machine*: photograph by Tom Jennings
144, *Shocking Duds*: left, photograph by Joseph Benner;
right, top and bottom photography by Laura MacCary
146, *Bringing Down the House*: photography by Becky Holter
148, *The Un-SUV*: top and bottom left, photography by Niel Benson,
bottom right, photograph by David Hess
150–153, *A Bug's Life*: photography by Frans Vandemaele
154, *Pedal Pusher*: photography by J.B. Straubel
156, *Home UFOs*: top, photograph by Jun Jalbuena; bottom, photograph by Virgil Polit
158, *Why Knot*: photograph by Katherine Frey, illustrations by Seth Goldstein
160, *Backyard-o-sphere*: photography by Sara Huston

162, *Flying from Scratch*: photograph and illustration courtesy Michel Therrien
164, *Brainy Bucket*: photograph by Ted Selker
165, *20-Horsepower Nut Grabber*: photograph by George Turnbull
166, *Microwave Ranger*: photography by Sigurd Kimpel
168, *Cocktails at 700 Mhz*: photography by Joel Wollborg
170, *Against the Grain*: top, photograph by Paul Moore;
bottom, photograph by Paul Taylor
172, *The Long Tail*: top, photograph by Mike Monaco;
bottom, photograph by Dan McKeone
174, *Speed Demon*: photograph by Ken Brough
176, *Hardwood Jets*: photograph courtesy Mark Nye
178, *Hardwood Jets*: photography by Mark Nye
179, *Hardwood Jets*: photograph courtesy Mark Nye

Haven't had enough?

Give the Maker in your life (or yourself) a truly unique gift — a subscription to MAKE magazine.

Receive $5 off a one-year subscription to MAKE.

Visit **makezine.com/subscribe** and use "G5MKER" as your promotional code.

*» Be sure to checkout the latest projects, creative mischief, ideas and free podcasts at **http://makezine.com***